Snorri Sturluson

Edda

Háttatal

Snorri Sturluson

Edda

Háttatal

Edited by
ANTHONY FAULKES

CLARENDON PRESS · OXFORD

OXFORD
UNIVERSITY PRESS

Great Clarendon Street, Oxford OX2 6DP
United Kingdom

Oxford University Press is a department of the University of Oxford.
It furthers the University's objective of excellence in research, scholarship,
and education by publishing worldwide. Oxford is a registered trade mark of
Oxford University Press in the UK and in certain other countries

First Edition published in 1991

Published in the United States of America by Oxford University Press
198 Madison Avenue, New York, NY 10016, United States of America

British Library Cataloguing in Publication Data
Data available

Library of Congress Cataloging in Publication Data
Data available

ISBN 978-0-19-811238-9

Contents

Manuscript Sigla

R Reykjavík, Stofnun Árna Magnússonar, Gks 2367 4to; ed. *SnE* (1931); facsimile in Wessén (1940).

T Utrecht, University Library, 1374; ed. W. van Eeden, *De Codex Trajectinus van de Snorra Edda* (Leiden, 1913); Árni Björnsson, *Snorra Edda* (Reykjavík, 1975); facsimile in *Codex Trajectinus*, ed. Anthony Faulkes (Early Icelandic Manuscripts in Facsimile, 15; Copenhagen, 1985).

U Uppsala, University Library, DG 11; ed. *SnE* 1848–87, ii. 250–396; facsimile in *Snorre Sturlas(s)ons Edda. Uppsala-handskriften DG 11* (i Stockholm, 1962; ii Uppsala, 1977).

W Copenhagen, Arnamagnæan Institute, AM 242 fol.; ed. *SnE* 1924; facsimile in *Codex Wormianus* (Corpus Codicum Islandicorum Medii Aevi, 2; Copenhagen, 1931).

Introduction

The Poem and its Author

Háttatal is an Icelandic poem in 102 stanzas divided into three sections (*kvæði*) which exemplifies a wide variety of verse-forms available to Norse poets in the thirteenth century, accompanied by a prose commentary that points out the main features of each verse-form. The content of the poem is praise in traditional skaldic style of Hákon Hákonarson, king of Norway 1217–1263, and his co-regent and future father-in-law Earl Skúli (1188/9–1240), for their generosity and valour in battle. The first section, stt. 1–30, is about Hákon, the second, stt. 31–67, is about Skúli, except for st. 67, which is about both rulers; in the third, stt. 68–95 are also mainly about Skúli, stt. 96–102 again seem to relate to both rulers; see note to stt. 1–30 and Möbius (1879–81), i. 35–6. It was composed by Snorri Sturluson (1179–1241), apparently in Iceland after his first visit to Norway to visit the two rulers in 1218–20, and is thus an example of the modification of the skaldic tradition brought about by the increasing use of the written word for literature in Scandinavia: instead of being recited aloud from memory by the poet in person before the ruler and his court, Snorri's poem was (presumably) sent in manuscript form to its patrons to be read by them or to them by someone else. There is no record of its reception. It is found in manuscripts as the third part of Snorri's *Edda*, after *Gylfaginning* (with its prologue) and *Skáldskaparmál*, though it is likely to have been the first part to have been composed (see Wessén 1940, 31–2).

Snorri's authorship of the poem is unequivocally confirmed both by attribution in the earliest manuscript (U) and by the attribution to him of extracts from it quoted in *Hákonar saga Hákonarsonar* and the third *Grammatical Treatise* (both by his nephew Sturla Þórðarson) and in additions to *Skáldskaparmál* in W (cf. also the reference in *TGT* 96: *í háttatali því er Snorri hefir ort*; the title is also found in rubrics in U and with the quotations from the poem in *Hákonar saga* and *TGT* and in the heading in T). Whether he also wrote the commentary, and whether this was sent to the rulers of Norway along with the poem, is much less certain. The prologue to the *Grammatical Treatises* in W seems to refer to it naming Snorri as the author (*eigi lengri reknar en Snorri lofar*, *GT Prologue* 155, cf. *Háttatal* 8/29–31), and the reference in *TGT* 96 can also be taken to refer to the commentary. But there are occasional discrepancies between the stanzas of the poem and the comments on them that cannot all be blamed on inaccurate copying, though they may be partly due to the poet's difficulty in analysing his practice if this was based on an intuitive rather than conscious grasp of principles. There are the following apparent examples of the commentary not fitting the verses:

(i) At 15/9–12 the etymology of the word *afleiðing* suggests that it belongs with the first word of st. 15 rather than the last of st. 14.

(ii) At 16/13 none of the possible meanings of *tíðar fall* seems to fit the pairs of words exemplifying *refhvǫrf* in stt. 17–22.

(iii) The similar phrase at 23/11, *í eina tíð*, also does not seem to fit the examples in st. 23.

(iv) Lines 23/11–16 seem to be in the wrong place, since st. 24 does not involve variation in length of line or in arrangement of hendings; these variations first appear in stt. 33 and 28 respectively.

(v) Lines 27/12–16 also seem to be in the wrong place, since the words would more naturally refer to the preceding verse, and it is rather artificial to make them introduce st. 28, with which they belong; cf. Möbius (1879–81), ii. 51–2.

(vi) Lines 28/12–13 mention the possibility of the *frumhending* falling on the second syllable of the lines with *skjálfhenda*, which happens in st. 33 but not in st. 28.

(vii) Lines 32/9–10 claim that the two hendings in the even lines both end *á einum hljóðstaf*, which is not the case.

(viii) Lines 77/9–10 state that there may be 7 syllables in a line of *hálfhnept*, but none of the lines of st. 77 has more than 6.

(ix) Line 79/10 states that the hendings in st. 79 are as in *dróttkvætt*, but l. 3 has *aðalhending*.

(x) The statement at 82/10–12, whatever the meaning of *hljóðstafr*, does not seem to apply to st. 82.

Some of these discrepancies may be oversights on the part of the author, especially if there was an interval between the composition of the poem and the compilation of the commentary. Others may be due to our misunderstanding of the author's terminology, which in some cases may have been experimental; and there is the perennial problem of medieval rationalization never quite conforming to our modern methods of analysis. Certainly the fact that Snorri is clearly able to handle resolution with short stressed syllables in his verse, though he is unable to give an explanation of it that conforms to Sievers's perception of it (6/22–8/14), should not be used as evidence that he was not the author of the commentary, though there do seem to be certain illogicalities and misunderstandings in the commentary that readers are reluctant to attribute to Snorri.

The kind of analysis and the way in which the dialogue form is handled (the speakers have no identity and the questions are not maintained throughout the work) are also different from what is found in *Gylfaginning* and *Skáldskaparmál*, though this may be because Snorri's techniques developed over the period he was engaged on his *Edda* (which may have been extensive, and *Háttatal* is assumed to have been written before the other two sections). The description of the *kenning* and particularly of the *sannkenning*, the meaning of the word *fornafn*, and the exemplification of *nýgjǫrvingar* are rather different in *Skáldskaparmál* from what we find in *Háttatal* (see under these words in the Glossary). In spite of all this there does not seem sufficient reason to doubt that the commentary is by Snorri, and that it formed part of his overall purpose in compiling the *Edda* as a handbook for young poets (see *Skáldskaparmál*, the so-called Epilogue, *SnE* (1931), 86/11–18 and cf. Kuhn (1983), 326). Though *Háttatal* is formally addressed to the rulers of Norway, its didactic purpose as an aid and encouragement to other poets and its intended Icelandic audience are obvious. There is a clear authorial voice in places in the commentary expressing unhesitant value-judgements on aspects of Norse poetry that there is no reason not to accept as Snorri's; these characteristically condemn excesses of various kinds, such as over-use of attributives (4/18–20), mixing of metaphors (6/16), repetition of words (8/27), using words in too extravagant senses (16/13–16), inconsistency in verse-forms (58/14–16), variation of metre for its own sake leading to lack of euphony (65/14).

There are clear references to identifiable historical events only in stt. 32–7, 39, and 63–6 (cf. Möbius 1879–81, i. 39), apart from the mentions of Snorri's visit to Norway in 1218–20 (e.g. st. 93; there are other references to his personal experiences in stt. 27–30, 67–70, 80–1, 95, 100–1). Stanzas 33–7 refer to the 'Vágsbrúarsumar' of 1214 (*Hákonar saga*, ch. 10; the saga, however, does not stress

Skúli's involvement in these events), and st. 32 refers to the killing of Páll *dróttseti*, recorded in the annals under 1214 (Storm 1888, 124, 183; *Flb* iv. 311), though the event probably took place late in 1213 (Munch 1857, 555). This killing is not mentioned in *Hákonar saga*, but it was in the longer version of *Bǫglunga sǫgur*, probably compiled *c.*1220 but now only extant in a Danish translation from about 1600 (*Bǫglunga sǫgur* 1988, ii. 126). It may be noted that although *Háttatal* gives the impression that Páll was killed in battle, he was in fact executed by Skúli for a treacherous attempt on the king's life. Stanza 39 mentions Skúli receiving an earldom from his half-brother King Ingi Bárðarson, which happened in 1217 shortly before the king's death (*Hákonar saga*, ch. 11); the two mentions in *Háttatal* of Skúli as *hertogi*, though, presumably only use the word in its general sense of 'war-leader', since Skúli only became duke in 1237 (*Hákonar saga*, ch. 190) and there is no likelihood that *Háttatal* was composed as late as that (cf. Möbius 1879–81, i. 33–4). Stanzas 63–6 refer to the events of 1221–2, when Skúli led an expedition against the Ribbungar in the Vík and a large party of them, including one of their leaders, Gunnarr Ásuson, was killed at Apaldrssetr (*Hákonar saga*, ch. 74; *Háttatal*, stt. 63–4 are quoted in this chapter and st. 66 in ch. 75, the only stanzas from *Háttatal* quoted in the saga). It is rather odd that st. 64 describes Gunnarr as fighting successfully and does not actually mention his death; perhaps Snorri received a garbled account of what happened. These events took place after Snorri's return to Iceland and could not have been known there before the summer of 1222, so that is the earliest date for the composition of *Háttatal*; there is no reason to think that it was composed long after that date either, since no subsequent events are mentioned in it, nọt even the important happenings of 1223; cf. Möbius (1879–81), i. 34, and note to st. 95 below; and Finnur Jónsson (1920–4), ii. 86. Konráð Gíslason (1869), 147–8 accordingly dated *Háttatal* to 1222–3, and this is the accepted date for the poem. Other references to the king's activities and dealings with his enemies are described in rather general terms in *Háttatal* that hardly allow identification with particular datable events (stt. 4–9, 17–22, 49–62; cf. Finnur Jónsson 1920–4, ii. 79); the references to the punishment of evil-doers (stt. 1, 17) could, however, refer to the expeditions against the Slittungar (*Hákonar saga*, chs. 47–8) and Ribbungar (*Hákonar saga*, chs. 66–9).

In spite of the mentions of the historical situation in Norway around the time of Snorri's first visit, there is very little evaluation on his part of the critical conflicts that were taking place and in which we may suppose he took a keen interest (there is no mention, for instance, of the conflicts between Norwegians and Icelanders or of the proposed military expedition against Iceland in 1220, *Hákonar saga*, ch. 59). Most of his comments are designed simply to glorify the position and qualities of Earl Skúli—sometimes to the implied detriment of the king himself. The over-prominence Snorri gives to Skúli, however, ironically foreshadows his adherence to the earl's side in the ensuing struggle for supremacy which the king eventually won, an adherence which was one of the factors leading to Snorri's death in Iceland at the hands of an emissary of the king in 1241. Snorri seems already in *Háttatal* to undervalue King Hákon, his praise often seeming even more perfunctory in his case than usual in skaldic verse; the king was after all too young at the time to have achieved the martial prominence that Snorri's verse attributes to him (he was born in 1204), and the eulogy comes dangerously close to being *háð en eigi lof* ('scorn rather than praise', *Hkr* i. 5). The historical references are therefore more informative about Snorri's own political leanings than about the actual events or situation in Norway when he visited it, though indeed the king's youth may be his main reason for devoting the majority of the poem to Skúli; there must have been rather little to say about Hákon at that time. Skúli was, moreover, in fact the principal ruler of Norway and at the zenith of his power at the time of Snorri's visit, having been granted a third of the country on his own account and being effectively regent of the rest because of Hákon's youth (*Hákonar saga*, chs. 22 and 54). He must have seemed to most people around 1220 the most important man in Norway and a candidate for future kingship (see Helle 1958, 92), even though he did not openly try to become king until somewhat later (1223, *Hákonar saga*, ch. 88). By the time of Snorri's second visit to Norway (1237–9) there was open rivalry and sometimes hostility between King Hákon and Earl Skúli, and Snorri not only spent most of his time with the latter, but also deliberately flouted the king's orders, showing clearly which ruler he preferred (*Hákonar saga*, ch. 195). Moreover, the anecdote in *Hákonar saga*, ch. 194 suggests that Skúli responded favourably to Snorri's poetic gifts; there is no evidence that King Hákon did. It is clear that on both his visits, Snorri was closer to Earl Skúli than to the king (*Sturl.* i. 271–2, 277–8, 444).

Similarly, Snorri's references to himself in the poem are little more than the standard skaldic references to the poet's art and to the hospitality and generosity of the recipients of his poem (stt. 27–31, 67–70, 80–1, 93, 95, 100). A few stanzas refer to his journey from Iceland to Norway and provide the opportunity for some not unskilful descriptions of sailing (stt. 27, 101; cf. also stt. 19–22, 34, 36, 38, and 71–9, which are apparently about expeditions in which Snorri took part while in Norway, and stt. 23–4, 29, 83, 86–9, and 91, which describe the feasting at the Norwegian court—Snorri really seems to have been somewhat dazzled by the spendour there), but it would be as much a mistake to look in this poem for expression of personal response to experience as it is to look for it in the poems of Snorri's predecessors. A major part of the poem is devoted to lively but unspecific and undetailed descriptions of battles supposedly engaged in by Hákon and Skúli. The poem is both a technical exercise and a part of the ritual of relationships between courtier and rulers, and is not the place for self-revelation.

Influences on the Work

The forerunners of *Háttatal* are of three kinds. Many of the verse-forms Snorri exemplifies had been used before by Norse poets (Norwegian or Icelandic) of the Viking Age or later, though in some cases the origins may have been in medieval Latin verse. A poem exemplifying various Norse metres known as *Háttalykill*, 'key of metres', had been composed by Hallr Þórarinsson and Earl Rǫgnvaldr of Orkney in the 1140s. And there existed a number of Latin treatises on metre with examples, as well as poems exemplifying the varieties of Latin metres.

In Latin literature probably the best-known treatment of metre was Servius *De centum metris* (Keil 1855–80, iv, pp. xlv–xlvii, 456–67). It contains a hundred examples of different metres composed by the author, who refers to the work as 'centimetrum' (p. 457). The Venerable Bede, in his own *De arte metrica*, ch. 24 (1975, 138), uses the phrase *in libris centimetrorum* as if such works were common. Aldhelm also wrote a treatise *De metris* (*Epistola ad Acircium*; 1919, 61–204). It is not certain that any of these Latin treatises were known directly in medieval Iceland. While *Háttatal* more than any other of the writings attributed to Snorri is reminiscent in manner and style and approach of the learned Latin treatises (particularly in its opening), the influence of any specific work cannot be demonstrated either on its form or its actual scheme of categorization and vocabulary. All that can be said is that the idea of composing a poem to exemplify (about) a hundred verse-forms with a commentary is very likely to have been suggested by knowledge, even if indirect, of one or more of the above works. As in other matters, Snorri seems to be influenced by foreign models, but his actual product is not imitative or derivative but based on firmly native tradition, though perhaps not for the most part on his own original perceptions and analysis.

There is remarkable correspondence between the opening of *Háttatal* and the beginning of Fortunatianus, *Ars rhetorica*, book 3 (4th c. AD; Halm (1863), 120–1, with the text corrected from Faral (1924), 55 n. 2):

Quot sunt generales modi dispositionis?
Duo.
Qui?
Naturalis et artificialis, id est utilitatis.
Quando naturalem ordinem sequemur?
Si nihil nobis oberit in causa.
Quid si aliquid occurrerit necessitate utilitatis?
Ordinem immutabimus naturalem.
Et quid sequemur?
Artificialem.
Quot modi sunt naturalis ordinis?
Octo.
Qui?

Totius orationis ⟨per partes⟩, per tempora, per incrementa, per status, per scriptorum partes atque verba, per confirmationis ac reprehensionis discrimen, per generales ac speciales quaestiones, per principales et incidentes.

But Fortunatianus is concerned with rhetoric, not metre, and his categories are different; only the manner is similar, and no direct influence is likely. Similarly, the use of the categories *setning*, *leyfi*, *fyrirboðning* seem to correspond to the three categories of Latin theorists (*pars praeceptiva*, *pars permissiva*, *pars prohibitiva*; see note to 0/7), but again the categories are applied in a different way and to metrics rather than grammar or rhetoric, and the similarity shows no more than a passing familiarity with the manner of Latin textbooks.

Háttalykill was probably not designed as a survey of available Norse metres, but was a *tour de force* involving the use of various metres, some of which were traditional Norse ones, but some of which were clearly modelled on poetry in other languages (on *Háttalykill* see Kuhn 1983, 317–19). As pointed out in JH–AH 120–1, of the metres represented in *Háttalykill*, only those of stt. 1 (*ljóðsháttr*), 2 (*kviðuháttr*), 3 (*dróttkvætt*), 8 (*munnvǫrp*), 26 (*háttlausa*), 13 (*tøgdrápulag*), 19 (*Bálkarlag*), 24 (*runhent*), 16 (*hrynhenda*) are clearly traditional Norse metres that were used at all widely in early verse. Snorri, in reproducing both the traditional and innovatory forms of *Háttalykill* as well as giving examples of other metres that had never in the ordinary way been used by Norse poets before, must also to a certain extent be said to be exemplifying what was theoretically possible in Norse verse, not what had actually been practised. (On use of foreign metres in *Háttatal* cf. Bjarni Einarsson 1969, 30; Heusler 1925, i. 313–14.) One untraditional feature in Snorri's practice in *Háttatal* is the tendency to end-stopping of lines, which is particularly noticeable (and indeed almost unavoidable) in the *runhent* metres and *hrynhenda*, which are dominant in the latter part of the poem. *Hrynhenda* also tends to trochaic metre, which emphasizes its similarity to Irish and Latin forms. A form of *hrynhenda* and *runhent* are combined in stt. 90–1; the result is nearly identical with the ecclesiastical trochaic tetrameter, which may indeed be the origin of this type (see JH–AH 131). *Dunhenda*, *iðurmælt*, and *klifat* use devices that are common 'colours' in classical rhetoric (JH–AH 127–8). *Stúfar*, of which Snorri has three variants (*Háttalykill* has only one, st. 31, where the b-lines are *stýft*) are rare outside these poems, and may also owe something to the influence of foreign metres (JH–AH 120). The device of *refrún/refhvarf*, not really common in Old Norse poetry (though it is used in *Háttalykill*, stt. 20, 28, 35, and there are seven kinds in *Háttatal*) may also originate in Latin tradition (JH–AH 124). The same may be true of *greppaminni*, *sextánmælt*, and others (JH–AH 125–7). There are also many verse-forms both in *Háttalykill* and *Háttatal* that are not found in earlier Norse verse except as sporadic variants in individual lines of verses that are otherwise normal *dróttkvætt*, and many of these may originate outside the Norse area too. Jón Helgason and Anne Holtsmark make much of the learned and foreign influences on the poets of *Háttalykill*; they point out that the twelfth century was a time of experimentation in European metres, and that there was a great deal of influence between Latin and the vernaculars in lyric forms.

But though the forms used in *Háttalykill* and *Háttatal* show wide variation, they include few forms that use anything other than the eight-line stanza (only *Háttalykill*, stt. 5 and 22, which each have ten lines; *Háttatal*, st. 100 and *Háttalykill*, st. 1, which have six lines; and *Háttatal*, st. 101, which has seven) and none that lack alliteration; rhyme is variable in its nature and placing, and not always present. Line-length varies from three syllables to nine; rhythms are almost infinitely variable, but very few of those used in *Háttatal* are not Sievers types (stt. 72, 76, and 84; see Sievers 1893, 113, 115). Kuhn (1983), 324 points out that fifty-six of the stanzas in *Háttatal* are basically *dróttkvætt* in form (as are nineteen of those in *Háttalykill*), namely stt. 1–32, 35–48, 52–8, 66–7, 88; the variations in stt. 24, 36–8, 41–8, 52–8, 66–7 concern only the arrangement of the hendings. Those of Snorri's forms that do not conform to the traditional rules of versification (according to Kuhn 1983, 325, stt. 9–10, 28, 37–8, 40, 43–4, 48, 57, 59–61) are rarely found used consistently in whole poems outside the *Háttalyklar*.

Only one of the verse-forms of *Háttalykill* is completely lacking in *Háttatal* (*núfuháttr*, st. 22). But many of the correspondences between verses in the two poems are only approximate, and *Háttatal* often divides *Háttalykill* forms into several variants (as for instance with *tøglag*), and often has greater regularity and organization in its forms as well as greater complexity; the order is made more rational. *Háttalykill* is of course incomplete, but fewer than half the forms in *Háttatal* correspond at all closely to those of the earlier poem as it survives, and about forty-five could be said to have no antecedent in it at all.

Háttalykill largely lacks the self-consciousness of *Háttatal*; the poets do not speak so much of their art in the poem as Snorri does in his (see, for example, *Háttatal*, stt. 95–7, 100, 102), and there is no commentary preserved. It also lacks the organization of the later poem. The verse-forms seem to appear in random order. It is not an exemplification of the court praise-poem; the subject-matter is an account of kings and heroes of the past, to a certain extent in chronological order, so that the poem belongs more with genealogical poems like *Ynglingatal* and *Háleygjatal* or catalogue-poems like *Íslendingadrápa* than with *Háttatal*.

The precedents for the verse-forms of *Háttatal* in Norse poetry are listed in the Appendix. From this it will be seen that there are clear precedents for just over thirty of them; just over thirty have no precedents at all except in some cases in *Háttalykill*; and the rest have partial precedents, that is to say the features of the verse-forms are found in individual lines of earlier verse, but not used consistently throughout a stanza or poem. Put another way, most of the variations of *dróttkvætt*—the stylistic or rhetorical ones, the syntactical ones, the use of elision, contraction, and resolution, those that involve particular arrangements of hendings and alliteration, and those that use particular rhythmical patterns, exemplified in stt. 2–48, 52–8, and 66–7—are found in earlier poems except for those of stt. 14 and 16, but except for stt. 13, 28, 35, and 66 generally only in odd lines and in some cases extremely rarely (as for instance *refhvarf*, stt. 17–23). There are also precedents for *hrynhenda*, *tøglag*, *hálfhnept*, *Haðarlag*; the *runhent* forms based on *fornyrðislag*, *málaháttr*, *dróttkvætt*, and *hrynhenda*; *kviðuháttr* and the 'eddic' metres in stt. 62–4, 68–70, 77, 79–81, 83, 85–8, 90–1, 94–102, though not generally conforming consistently to any one of the subtypes that Snorri exemplifies. But there are no real precedents at all except in *Háttalykill* for the *stúfar* (except in so far as these coincide with *hálfhnept*), *kimblabǫnd*, *draughent*, *grœnlenzki háttr*, *nýi háttr*, *stúfhent*, *náhent*, *hnugghent*, and *alhnept* of stt. 49–51, 59–61, 65, 71–6, 78, or for the varieties of *runhent* in stt. 82, 84, 89, 92–3.

Háttatal exemplifies the development in attitudes to poetry in Iceland that led both to a greater awareness of literary and linguistic theory and to more academic styles of composition. *Háttalykill* shows that this development was well under way by the middle of the twelfth century; it is also already clearly discernible in the poetry of Einarr Skúlason, a great experimenter both with form and content, as well as a careful observer of the 'rules' (cf. Kuhn 1983, 312–17). *Alvíssmál* and the *þulur* show the same underlying trend (whenever it was that they were composed), and Haukr Valdísarson's *Íslendingadrápa*, *Málsháttakvæði*, and Bjarni Kolbeinsson's *Jómsvíkingadrápa* can be considered to reveal the same academic tendency. These last three poems are all preserved in manuscripts of Snorri's *Edda*, like the four *Grammatical Treatises*, which display the more specifically analytical attitude that developed towards both language in general and poetry in particular once literacy was established in Iceland. Several of the poems mentioned above have connections with Orkney, and there were also close relations between Orkney and Oddi in southern Iceland, where Snorri was brought up. It has seemed likely to many scholars that there might have been a 'school' of (or indeed for) poets at Oddi with influence from and on Orkney. Both places seem to have been particularly open to influence from more southern parts of Europe while at the same time being especially concerned to preserve knowledge of native Scandinavian tradition and culture (see Bjarni Guðnason 1963, 258 n.; on Orkney as an intermediary between the cultures of the European continent and Iceland see Chesnutt 1968). There is however little actual evidence for the existence of formal training of vernacular poets in medieval Iceland, though if there had been, Snorri's *Edda* would have been the ideal textbook (cf. Paasche 1957, 411–12); all the sources indicate is informal apprenticeship such as is described in *Egils saga*, ch. 78 (*ÍF* ii. 268). Composition of *Latin* verse was evidently taught at the cathedral school at Hólar; see *Biskupa sögur* (1858–78), i. 235–41.

Influence on Later Writers

As a composer of praise-poetry for the Norwegian court, Snorri is almost at the end of the tradition, and his aim of revitalizing the activity was not realized. There was indeed another generation of poets who composed in honour of the kings of Norway after him, of whom the most distinguished was his nephew Sturla Þórðarson, but in the thirteenth century praise-poetry had become a largely literary activity (*Háttatal* itself, if it was delivered at all, must have been delivered in manuscript form), and it is doubtful whether any new court poetry was transmitted or preserved orally after Snorri's time (Sturla's poems are preserved as quotations, presented as sources, in his own prose sagas). But as a means of preserving the memory of historical events, as well as as an organ of royal propaganda, skaldic poetry was being superseded by the written prose saga—had indeed been since the time of King Sverrir—and as a part of the ritual and entertainment of the court was being superseded by various kinds of prose narrative, including translated romances; taste in poetry was moving to favour the ballad and its derivatives; in Iceland a new genre, the *rímur*, was to replace skaldic verse as a medium of entertainment both written and oral. The skaldic tradition continued for a generation or so in altered style as a medium for religious poetry, both narrative and devotional, and some fourteenth-century poets refer to *eddu reglur* and *eddu list* ('the rules of [the] *edda*', 'the art of [the] *edda*'; *Skj A* ii. 348, 394, 429). The references are usually part of a humility *topos* (the writer claims he knows nothing about the rules and art of poetry), and *edda* by now may mean nothing more than 'poetics' or 'rhetoric' in general, but the remarks are evidence that as a result of Snorri's work there was a greater awareness of poetic theory in late medieval Icelandic poets. For these religious poets as for the composers of *rímur* it is likely that *Skáldskaparmál* was more influential than *Háttatal*, though the latter probably played its part in encouraging the metrical experimentation, variation, and complexity that characterizes Icelandic poetry in the late Middle Ages and later. The use of unusual metres in verses quoted in *Sturlunga saga* is also probably due in part to the influence of *Háttatal*; some of them seem to be direct imitations of forms in Snorri's poem (see e.g. JH–AH 126); Snorri's nephew Sturla Þórðarson also clearly chose metres for his poems from those exemplified in *Háttatal*. The specific tradition of the 'key of metres' initiated by Rǫgnvaldr Kali, Hallr Þórarinsson, and Snorri was continued by various Icelandic poets in the late Middle Ages and afterwards: for example, the *Háttalyklar* attributed to Loptr inn ríki Gutthormsson (*Småstykker* 1884–91, 203–82, 297–344; see Finnur Jónsson 1920–4, iii. 20–2); *Máríulykill* (Jón Helgason 1936–8, ii. 203–28); the *Háttalykill* of Þórður Magnússon á Strjúgi (*Småstykker* 1884–91, 345–60); and that of Hallur Magnússon (see Jón Þorkelsson 1888, 361–7, cf. 243 n. 1; Möbius 1879–81, i. 43 n.).

Snorri's technical vocabulary is different from that of *FGT*, and overlaps to only a limited extent with that of *SGT* (see *hljóðstafr* in Glossary; compare *SGT* 50–4 with the beginning of *Háttatal*) and *TGT* (see, for example, *samstafa* in Glossary), though *TGT* does quote and use *Háttatal* in various places. It is *FoGT* that is most clearly influenced by Snorri's work, quoting it quite extensively (see especially *FoGT* 136–7) and using much of its terminology. The prologue to the *Grammatical Treatises* in W (probably written by the compiler of *FoGT*: Finnur Jónsson 1920–4, ii. 922 n. 3) also refers to *Háttatal* (as well as to *Gylfaginning* and *Skáldskaparmál*). The treatises, however, are in general more learned than Snorri, and show clearer signs both in their terminology and methods of knowledge of particular Latin treatises, *TGT* and *FoGT* being based quite closely on extant Latin sources, principally Priscian, Donatus, and Alexander de Villedieu; see Foote (1982); Raschellà (1983), 302–4.

The names Snorri (or later scribes of his work) gave to the various verse-forms of *Háttatal* may in some cases have been traditional before his time (a number appear in headings in *Háttalykill*, though often in variant forms which may not be purely scribal), but a number of them were obviously originally names for phenomena occurring sporadically in skaldic verse (such as *stælt*, *hjástælt*, *tiltekit*, *tilsagt*, *refhvarf*, *þríhent*, *dunhenda*, *náhendr*), and only when Snorri had constructed stanzas using them consistently did they become names of verse-forms (Kuhn 1983, 326). As a result of the influence of *Háttatal* many of them have remained in use in modern Icelandic (about 40% of Snorri's terms appear in Helgi Sigurðsson 1891, 13–38, though not always with the same meanings as in *Háttatal*).

Evaluation

Though Snorri is now celebrated mainly for his prose writing, particularly *Heimskringla*, but also for his mythological work in *Gylfaginning* and *Skáldskaparmál*, his contemporaries seem to have thought more highly of him as a poet. While there is only passing reference to his historical works in *Sturlunga saga* (*Sturl.* i. 342), he is in several places referred to as a poet (*Sturl.* i. 269, 271, 278, 284; cf. *Gunnlaugs saga*, *ÍF* iii. 51 n. 3; see Bjarni Einarsson 1969, 27), though without special reference to *Háttatal*; but evaluation of his other work is difficult when so little of it survives: *Skáldatal* (*SnE* 1848–87, iii. 270–86) mentions poems now lost on King Sverrir, King Ingi Bárðarson, Earl Hákon galinn (d. 1214; see *Sturl.* i. 269); a poem *Andvaka* on Earl Hákon's wife is mentioned in *Sturl.* i. 271 (see Bjarni Einarsson 1969); of two poems on Earl Skúli besides *Háttatal*, all that is preserved is the *stef*, three lines, of one of them, *Sturl.* i. 278, cf. 284 and *Háttatal* 69; of a poem apparently about Bishop Guðmundr Arason only a couplet survives (quoted *TGT* 76). Snorri was evidently quite a prolific poet, but besides *Háttatal* and the two fragments just mentioned all that survives of his poetry is six and a half *lausavísur* quoted in *Sturlunga saga*,

TGT, *FoGT*, additions to *Skáldskaparmál* in W, and *Hákonar saga Hákonarsonar*.

Scholars this century have not rated *Háttatal* highly as a poem: the usual judgement has been that it shows technical brilliance but lacks inspiration—a more extreme version of the commonest view of skaldic praise-poems in general; cf. Finnur Jónsson 1920–4, ii. 88, who rates the *lausavísur* higher; Jón Helgason (1953), 131, 157–8; Hallberg (1975), 2. Spontaneity, expressiveness, and sincerity were probably not, however, qualities thought indispensable in court poetry in the Middle Ages, when artistry and control of technique were more highly valued. Hans Kuhn, indeed, even criticized Snorri's versification, and has claimed that he reveals that he does not understand certain aspects of Norse metre (1983, 327; 1969*a*, 230–1), and that he fails to observe some of the restrictions on placing of parts of speech in the verse-line that were almost invariably observed by earlier skalds, constructing, for instance, lines with a nominal element before the first stave (1969, 226; cf. 1983, 282, 325–7). In fact various changes in the language (e.g. changes in the relative stressing of different word-classes and blurring of quantitative distinctions, see Noreen (1923), § 125) resulted in poets of the later Middle Ages perceiving stress and metre rather differently from poets of the Viking Age, and it was becoming impossible to preserve the nuances of line-structure that had been natural to those poets in any case. *Háttatal* also seems more often to use artificial word-order than earlier skaldic verse, though not so frequently if the interpretations proposed by E. A. Kock in *NN* are accepted, rather than those of Finnur Jónsson in *Skj* B and *LP*; but the use of complex, almost regularly interwoven clauses at times (e.g. in stt. 36, 78, 98) becomes reminiscent of *vers rapporté* (on which see Raynaud de Lage 1951, 157–8 and de Vries (1964–7), ii. 31; cf. *Háttalykill*, st. 35). Snorri's lapses are not unprecedented, though they were usually avoided by the 'stricter' court poets, and various kinds of licence seem to have been permitted in the less formal verse such as is often quoted in Sagas of Icelanders; see Kuhn (1983), 218–20; (1937), 63, where he quotes references to such freedom in Sighvatr's *lausavísa* 29 (*Skj A* i. 274), Bjǫrn Hítdœlakappi's *lausavísa* 21 (*ÍF* iii. 190), and Sneglu-Halli's *lausavísa* 7 (*ÍF* ix. 292–3); cf. also Kuhn (1981). The only stanzas in *Háttatal* that seem not to be analyzable according to Sievers's principles seem to be 72 (the even lines), 76/1, 3, 5, and 84. It might be said that Snorri's use of kennings shows little imagination too, considering the range of knowledge of them he shows in *Skáldskaparmál*, but a spirited defence of his verbal artistry has been made by Bjarne Fidjestøl (1982), 246–55. One apparent lack in a poem that is supposed to be a model for court poets is that the poem overall lacks a *stef*; but this is a consequence of varying the metre from stanza to stanza, making the traditional kind of *stef* impossible (the only use of a *stef* is the *klofastef* exemplified in the section on *tøglag*, stt. 68–70; cf. Fidjestøl 1982, 248). Though many will still find the content of *Háttatal* rather repetitious and tedious, most will be able to respond to the poet's exultation and pride in his technical originality and achievement in his claim to have composed praise-poetry of an unprecedented kind and that his poem, his *tour de force*, will be remembered for ever (stt. 70, 95–7, 100, 102).

The commentary is a remarkable achievement; the level of analysis of the more technical aspects of verse-structure is impressive. But in a number of places the writer fails to bring out completely what seem to be the real distinctive features of stanzas (cf. Kuhn 1983, 326). In particular he clearly has only an intuitive grasp of resolution (the substitution of a disyllable with short first syllable for a long syllable in stressed position) and is unable to describe it accurately (see, for instance, 7/11–8/14). In general, though, if the writer does not analyse the metres quite in the way that a modern writer would, he shows remarkable perception given the means available to him. Another feature that disturbs modern analysts is that he tends to assign the name of a class of phenomena to a subclass of itself (e.g. 2/10–11 and 4/21–2), but this is common medieval (and indeed classical) practice, not a sign of incompetence.

In the arrangement of verse-forms, Snorri puts *dróttkvætt* first and *fornyrðislag* and its variants last, in defiance of the usual assumptions about the historical development of Norse metres. This, however is not due to ignorance. Snorri knew that eddic poems—and hence their metres—had a longer history than skaldic poems, as is evident from *Gylfaginning*, where he excluded the latter from within the dialogue to avoid anachronism (eddic poems he claims to have in some cases been composed by the Æsir in prehistoric times; skaldic poems are by historical vikings). The arrangement in *Háttatal* is not intended to be historical—which seems the natural approach only to modern scholars—but typological. The verse-forms are arranged and analyzed structurally and evaluated hierarchically, with the standard form (for praise-poems, which is

Snorri's topic) first and the simplest types, *inir smæri hættir*, those hardly, in Snorri's view, being elaborate enough for court poetry, at the end. The section on *fornskálda hættir* (stt. 54–8) shows that he was well aware of the way in which the historical development of *dróttkvætt* took place, and that he realized that variation in the arrangement of hendings was characteristic of the earliest skaldic verse, even if the precise variations he attributes to particular early poets are not specific to the work of theirs that has survived. His view of the status of the 'older' poets probably owes something to the medieval attitude to the classics.

Snorri's is by far the earliest medieval treatise on the metres of poetry in a Germanic language; before his time the only European vernacular in which such treatises were written was Irish (see Murphy 1961; de Vries (1964–7), ii. 32). His poem and its commentary are therefore of immense importance for our understanding of Germanic metres and of the ways in which medieval poets in northern Europe perceived their work. It is particularly remarkable that Snorri was contemporary with the first writers in southern Europe to deal in the vernacular with vernacular versification (Raimon Vidal, *c.*1200; Uc Faidit *c.*1240; see Patterson 1935, i. 34), though his work is not very similar to theirs (it is more similar in method and approach to the Irish treatises, but it is improbable that he could have been influenced by them either). It is possible that the Icelandic interest in poetic theory was one of the effects stimulated by the visit of Rǫgnvaldr Kali and his party to Provence in the mid-twelfth century, and by the subsequent mediation of cultural influence from Europe to Iceland via Orkney (cf. Heusler 1925, i. 313).

Preservation

There are four independent manuscripts that contain all or a substantial part of *Háttatal*. In the Codex Regius (R = Reykjavík, Stofnun Árna Magnússonar GkS 2367 4to, written about the middle of the first half of the fourteenth century) it is virtually complete, though some passages at the tops of pages have been rubbed so as to be nearly illegible (of two lines in one stanza, no. 94, very little can be read, and the text at this point is not extant elsewhere) and it seems clear that some words and phrases have been omitted from the commentary (e.g. that to st. 53); the commentary is entirely lacking for st. 38, which is placed at the end of the poem. The commentary in various places requires emendation to make sense. The text of the verses has in many places been corrected or amended by a second contemporary hand (R*), which has also added some headings (see Finnur Jónsson 1892). Codex Trajectinus (T = Utrecht, University Library 1374, written about 1595, but probably a copy of a thirteenth-century manuscript now lost) has a text very similar to that in R, though inaccurate in many places and ending after st. 61 (the last leaves of the manuscript are lost). Codex Wormianus (W = Copenhagen, Arnamagnæan Institute, AM 242 fol., written about the middle of the fourteenth century) included *Háttatal* after the four *Grammatical Treatises*, but the leaves containing stt. 1–6 and 87–102 are now lost (they have been replaced on paper by text copied in the seventeenth century from extant manuscripts) and various passages in the commentary are omitted. The Uppsala manuscript (U = Uppsala, University Library, DG 11, written in the early fourteenth century, and probably the oldest extant manuscript) has a text that breaks off after st. 56 (it seems that for some reason the scribe did not write any more) and is, as elsewhere in the prose *Edda*, rather different from the other manuscripts. The commentary seems in some places to have been shortened and is frequently incoherent, but also often contains words, phrases, and headings lacking in the other manuscripts. Although the text is often inaccurate, it may well be derived from Snorri's original independently of the hyparchetype of R, T, and W, or may even derive from an early draft made by Snorri. As well as the text down to st. 56, U has, before the heading to the poem (and after the version of the second *Grammatical Treatise* that is included after *Skáldskaparmál*) a list of the names of most of the first thirty-six verse-forms together with the opening lines of the corresponding stanzas (this part of the text is denoted U*). Since many of the names of the verse-forms are omitted from the other manuscripts, this is a most welcome addition to the text, whether or not the names derive from Snorri. It is, however, difficult to see any possible purpose in this arrangement of the text other than as an *aide-mémoire* to someone who knew the text of the poem by heart, but wanted to be reminded of the order of the verses and of the names of the verse-forms. It may have been used either in conjunction with performance, or, perhaps more likely, in conjunction with an oral discussion or lecture on the various metres represented. The reason for stopping with st. 36 is not apparent. Other material in the Uppsala manuscript (*Skáldatal*, the Sturlung genealogy, the list of lawspeakers) suggests that the manuscript is derived from a compilation made from Snorri's working papers.

Theoretically, agreement between the text of U and one other manuscript ought to represent the reading of the archetype (if U goes back independently to Snorri's original, this would have been an authorial version), but contamination and scribal revision make the situation less straightforward. In some places all four manuscripts seem to share errors (e.g. in the placing of 8/38–40). The oldest manuscript (U) was written more than half a century after Snorri's death and all four contain many patent inaccuracies and omissions. The text of *Háttatal*, therefore, cannot be said to be well preserved, and reconstruction of the original is scarcely possible, at least as regards the commentary; there is not often reason to doubt the essential accuracy of the text of the poem itself.

Stanzas 63, 64, and 66 are quoted in Sturla Þórðarson's *Hákonar saga Hákonarsonar* (chs. 74–5; with attribution to Snorri), stt. 5/3–6, 15/7–16/1, 28/3–4, 40/1–4, 73/1–4, 83/5–6 in the third *Grammatical Treatise* (with attribution to Snorri except in the case of 40/1–4; the first of these in AM 748 I 4to, the second, third, fourth, and fifth in AM 748 I 4to and (lines 1–2 only of st. 73) W, the sixth in AM 757 a 4to and W), stt. 12 and 14 in the fourth *Grammatical Treatise* (in W with no attribution), stt. 2/5–8 and 40/7–8 in the additions to *Skáldskaparmál* in W (with attribution to Snorri).

This Edition

The spelling is normalized to a thirteenth-century standard and abbreviations are expanded, without notice except where there could be ambiguity. Punctuation, capitalization, and paragraphing are added, though the occurrence of large capitals in R is noted.

Reconstruction of the author's original or of the archetype have both been judged impossible, and the text is based on R, supplemented where necessary (where the text does not give acceptable sense or is clearly damaged) from T, W, and U. Additions and corrections in the second hand in R when noted are marked R*, the names of metres and readings from the list of opening lines of stanzas in U are marked U*. Letters and words no longer legible in R are enclosed in square brackets, letters and words lacking in R and added from other MSS or by editorial conjecture are enclosed in pointed brackets. (Round brackets are part of the editorial punctuation.) An asterisk before a word indicates that it has been emended, the dagger-sign (†) that the order of words in R has been departed from, or that words in R have been relegated to the textual notes. The text has been checked with the manuscript, but some readings are legible in photographs made before its restoration that are no longer visible in the manuscript itself; and in some places the facsimile edition (1940) is more legible than either. Many of the readings reported in *SnE* (1931) and *Skj A* ii, however, cannot be substantiated, and these editions have too many clear inaccuracies of detail for such readings to be accepted. The textual notes record all departures from the text of R and the manuscript sources, if any, of emendations, but do not attempt in general to record variants, for which the reader is directed to the separate editions of T, W, and U.

Various rubricated headings in R are now indistinct or illegible (often it is even difficult to see whether there was a heading or not) and they are inconsistent in the other manuscripts. Where these rubrics of R can be deciphered they are given in the textual notes. There is sporadic numbering of the stanzas, sometimes rubricated (and difficult to decipher), and presumably by the original scribe, sometimes in black in the margin, and presumably the latter are later additions. The numbering is in some cases of verse-forms rather than stanzas, so that st. 9 is made no. 2 etc. (cf. 67/10–12 and note).

The glossary includes all technical and poetical words with full references, but many ordinary words that should cause no difficulty are omitted.

Further Reading

The fullest edition of *Háttatal* with discussion and commentary is Möbius (1879–81). There are notes and especially good interpretations of the verses (in modern Icelandic) by Magnús Finnbogason in *SnE* (1952). The only work in English is Martin (1974), which includes a translation. The text of course appears in the editions of

SnE, of which *SnE* (1931) has the fullest textual notes, though they are not entirely accurate, and in the facsimiles of the manuscripts (see p. vii above). There is a Latin translation in *SnE* (1848–87), i and an English one in Snorri Sturluson, *Edda*, tr. Anthony Faulkes (Dent: Everyman's Library, London, 1987). The text of the poem (and of other verses by Snorri) appears with Danish translation in *Skj* and there are comments on individual stanzas in *NN*. Metrical matters are discussed in Kuhn (1983) and there is literary discussion in Fidjestøl (1982) and Hallberg (1975), 2. The most important text for comparison with *Háttatal* is *Háttalykill enn forni* (1941). For Snorri's life and other writings see *Gylfaginning* (1982), pp. xxxiii–xxxiv, and further *Snorri: Átta alda minning* (Reykjavík, 1979); John Simon, 'Snorri Sturluson: His Life and Times', *Parergon*, 15 (1976), 3–15.

Snorri Sturluson

Edda

PART III

[Háttatal]
⟨er Snorri Sturluson orti um Hákon konung ok Skúla hertoga⟩

Hvat eru hættir skáldskapar?

Þrent.

Hverir?

Setning, leyfi, fyrirboðning.

Hvat er setning háttanna?

Tvent.

Hver?

Rétt ok breytt.

Hvernig er rétt setning háttanna?

Tvenn.

Hver?

Tala ok *grein.

Hvat er tala setningar háttanna?

Þrenn.

Hver?

Sú er ein tala, hversu margir hættir hafa fundizk í kveðskap hǫfuðskálda. Ǫnnur tala er þat, hversu mǫrg vísuorð standa í einu eyrindi í hverjum hætti. In þriðja tala er sú, hversu margar samstǫfur eru settar í hvert vísuorð í hverjum hætti.

Hver er grein setningar *háttanna?

Tvenn.

Hver?

Málsgrein ok hljóðsgrein.

Hvat er málsgrein?

Stafasetning greinir mál allt, en hljóð greinir þat at hafa samstǫfur langar eða skammar, harðar eða linar, ok þat er setning hljóðsgreina er vér kǫllum hendingar, svá sem hér er kveðit:

Lætr sár Hákun heitir 1
(hann rekkir lið) bannat
(jǫrð kann frelsa) fyrðum
friðrofs konungr ofsa;

sjálfr ræðr allt ok Elfar,
ungr stillir sá, milli
(gramr *á gipt at fremri)
Gandvíkr jǫfurr landi.

Hér er stafasetning sú er hætti ræðr ok kveðandi gerir, þat eru tólf stafir í eyrindi, ok eru þrír settir í hvern fjórðung. Í hverjum fjórðungi eru tvau vísuorð. Hverju vísuorði fylgja sex samstǫfur. Í ǫðru vísuorði er settr sá stafr fyrst í vísuorðinu er vér kǫllum hǫfuðstaf. Sá stafr ræðr kveðandi. En í fyrsta vísuorði mun sá stafr finnast tysvar standa fyrir samstǫfun. Þá stafi kǫllum vér stuðla. Ef hǫfuðstafr er samhljóðandi, þá skulu stuðlar vera enn inn sami stafr, svá sem hér er:

Lætr sá er Hákun heitir
hann rekkir lið bannat.

En rangt er ef þessir stafir standa fyrir samstǫfun optar eða sjaldnar en svá í fjórðungi vísu. En ef hljóðstafr er hǫfuðstafrinn, þá skulu stuðlar vera ok hljóðstafir, ok er fegra at sinn hljóðstafr sé hverr þeira. Þá má ok hlýða at hljóðstafr standi fyrir optar í fjórðungi í fornǫfnum eða í málfylling þeiri er svá kveðr at: 'ek', eða svá: 'en, er, at, í, *á, of, af, um', ok er þat leyfi en eigi rétt setning.

Ǫnnur stafasetning er sú er fylgir setning hljóðs þess er hátt gerir ok kveðandi. Skal sú grein í dróttkvæðum hætti svá vera at fjórðungr vísu skal þar saman fara at allri stafasetning ok hljóða. Skal í fyrra vísuorði þannig greina þá setning:

Jǫrð kann frelsa fyrðum.

Hér er svá: 'jǫrð . . . fyrð-'. Þat er ein samstafa í hvárum stað ok sinn hljóðstafr fylgir hvárri ok svá upphafsstaf[r en einir stafir eru] eptir hljóðstaf í báðum orðum. Þessa setning hljóðfalls kǫllum vér skothending. En í ǫðru vísuorði er svá:

Friðrofs konungr ofsa.

Svá er hér: '-rofs . . . ofs-'. Þar er einn hljóðstafr ok svá allir þeir er eptir fara í báðum orðum, en upphafsstafir greina orðin. Þetta heita aðalhendingar. Svá skal hendingar setja í dróttkvæðum hætti at hin síðari hending í hverju vísuorði, er heitir viðrhending, hon skal standa í þeiri samstǫfu er ein er síðar, en sú hending er frumhending heitir stendr stundum í upphafi orðs—kǫllum vér þá oddhending—stundum í miðju orði—kǫllum vér þá hluthending. Þetta er dróttkvæðr háttr. Með þeima hætti er flest ort þat er vandat er. Þessi er upphaf allra hátta sem málrúnar eru fyrir ǫðrum rúnum.

Hvernig er breytt setning háttanna?

Tvá vega.

Hvernig?

Með máli ok hljóðum.

Hvernig skal með máli skipta?

Tvá vega.

Hvernig?

Halda eða skipta háttunum.

Hvern⟨i⟩g skal breyta háttunum ok halda sama hætti?

Svá: at kenna eða styðja eða reka eða sannkenna eða yrkja at nýgjǫrvingum.

Hvat eru kendir hættir?

Svá sem þetta:

Fellr of fúra stilli **2**
fleinbraks, limu axla,
Hamðis *fang, þar er hringum
hylr ættstuðill skylja;
holt felr hildigelti
heila bœs, ⟨ok⟩ deilir
gulls í gelmis stalli
gunnseið skǫrungr reiðir.

Hér eru ǫll heiti kend í þessi vísu, en hendingar ok orðalengð ok stafaskipti fara sem fyrr var ritat. Kenningar eru með þrennum háttum greindar: fyrst heita *kenningar, annat tvíkent, þriðja rekit. Þat er kenning at kalla fleinbrak orrostu, en þat er tvíkent at kalla fleinbraks ⟨fúr⟩ sverðit, en þá er rekit ef lengra er.

Úlfs bága verr ægis **3**
ítr báls hati málu;
sett eru bǫrð fyrir bratta
brún Míms vinar rúnu;
orms váða kann eiðu
allvaldr gǫfugr halda;
menstríðir njót móður
mellu dólgs til elli.

Hvat eru sannkenningar?

Svá sem þetta:

4 Stinn sár þróask stórum,
sterk egg frǫmum seggjum
hvast skerr hlífar traustar;
hár gramr lifir framla.
Hrein sverð litar harða
hverr drengr; gǫfugr þengill
(ítr rǫnd furask undrum)
unir bjart⟨r⟩ snǫru hjarta.

Þat er sannkenning at styðja svá orðit með sǫnnu efni, svá at kalla stinn sárin, þvíat hǫfug eru sár stór; en rétt er mælt at þróask. Ǫnnur sannkenning er sú at sárin þróask stórum. Nú er eitt vísuorð ok tvær sannkenningar. Í ǫðru vísuorði er kǫlluð sterk egg, en framir seggir. Í inu þriðja er svá, at hvast skerr, hlífin er traust; ok í fjórða orði at kalla konunginn mikinn, en líf hans framligt, þar næst at kalla hreint sverð ok harðliga roðit, en einnhverr liðsmanna, ok væri rétt mál þótt maðr væri nefndr. Gǫfugr er konungrinn kallaðr, rǫndi⟨n⟩ var kostig ok furaðisk undarliga skjótt; konungrinn unði glaðr frœknu hjarta. Nú eru hér sýndar sextán sann[kenningar í átta] vísuorðum, en þó fegra þær mjǫk í kveðandi at eigi sé svá vandliga eptir þeim farit.

Sannkenningar hafa þrenna grein: heitir ein sannkenning, ǫnnur stuðning, þriðja tvíriðit.†

5 Óðharða spyr ek eyða
egg fullhvǫtum seggjum;
dáðrǫkkum veldr dauða
dreng ofrhugaðr þengill;
hamdøkkum fær Hlakkar
hauk munnroða aukinn
(*veghrœsinn spyr ek vísa)
vald⟨r⟩ ógnþorinn skjaldar.

Hér fylgir stuðning hverri sannkenning, svá sem kǫlluð er eggin óðhǫrð, en fullhvatir menninir. Þat er sannkenning: hǫrð egg, en hvatir menn. Þat er stuðning er annat sǫnnunarorð fylgir sannkenning.

Hvat eru nýgjǫrvingar?

Svá sem þetta:

6 Sviðr lætr sóknar naðra
slíðrbraut jǫfurr skríða;

ótt ferr rógs ór réttum
ramsnákr fetilhamsi;
linnr kná sverða sennu
sveita bekks at leita;
ormr þyrr vals *at varmri
víggjǫll sefa stígu.

Þat eru nýgjǫrvingar at kalla sverðit orm ok kenna rétt, en slíðrirnar gǫtur hans, en fetlana ok umgjǫrð hams hans. Þat heldr til ormsins nátturu at hann skríðr ór hamsi svá at hann skríðr mjǫk til vatns. Hér er svá sett nýgjǫrving at hann ferr leita blóðs bekkjar at þar er hann skríðr hugar stígu, þat eru brjóst manna. Þá þykkja nýgjǫrvingar vel kveðnar ef þat mál er upp er tekit haldi of alla vísulengð. *En *ef sverð *er ormr kallaðr, ⟨en síðan⟩ fiskr eða vǫndr eða annan veg breytt, þat kalla menn nykrat, ok þykkir þat spilla.

Nú er *dróttkveðinn *háttr með fimm greinum, ok er þó hinn sami háttr réttr ok óbrugðinn, ok er optliga þessar greinir sumar eða allar í einni vísu ok er þat rétt, þvíat kenningar auka orðfjǫlða, sannkenningar fegra ok fylla mál, nýgjǫrvingar sýna kunnustu ok orðfimi.

Þat er leyfi háttanna at hafa samstǫfur seinar eða skjótar, svá at dragisk fram eða aptr ór réttri tǫlu setningar, ok megu finnask svá seinar at fimm samstǫfur sé í ǫðru ok inu fjórða vísuorði, svá sem hér er:

Hjálms fylli spekr hilmir **7**
hvatr Vindhlés skatna;
*hann kná hjǫrvi þunnum
hræs þjóðár ræsa;
ýgr hilmir *lætr eiga
ǫld dreyrfá skjǫldu;
styrs rýðr stillir hersum
sterkr járngrá serki.

Í þessi vísu eru allar oddhendingar ⟨inar fyrri hendingar⟩, ok er þó þessi háttr dróttkvæðr at hætti.

Nú skal sýna svá skjótar samstǫfur ok svá settar nær hverja annarri at af því eykr lengð orðsins:

Klofinn spyr ek hjálm fyrir hilmis **8**
hjara⟨r⟩ egg; duga seggir;

því eru heldr þar er sk⟨e⟩kr skjǫldu
skafin sverð lituð ferðar;
bila muna gramr þó at gumna
gular rítr nái líta;
draga þorir hann yfir hreina
hvatan brand þrǫmu randa.

Hér er í fyrsta ok þriðja vísuorði níu samstǫfur, en í ǫðru ok í fjórða sjau. Hér er þat sýnt hversu flestar samstǫfur megu vera í vísuorði með dróttkvæðum hætti, ok af þessu má þat vita at átta eða sjau megu vel hlýða í fyrsta ok þriðja vísuorði. Í þessi vísu eru allar frumhendingar hluthendur, ok dregr þat til at lengja má orðit, at sem flestar samstǫfur standi fyrir hendingar.

Þat er annat leyfi háttanna at hafa í dróttkvæðum hætti eitt orð eða tvau í vísu með álǫgum eða detthent eða dunhent eða skjálfhent eða með nokkvorum þeim hætti er eigi spilli kveðandi. Þriðja leyfi er þat at hafa aðalhendingar í fyrsta eða þriðja vísuorði. Fjórða leyfi er þat at skemma svá samstǫfur at gera eina ór tveim ok taka ór annarri hljóðstaf. Þat kǫllum vér bragarmál, svá sem hér, er kvað Þórarinn máhlíðingr:

*Varðak mik þars myrðir
morðfárs vega *þorði.

Enn er sú grein út sett myklu lengra. Þat er hit fimta leyfi at skipta tíðum í vísuhelmingi. Sétta leyfi er þat at hafa í dróttkvæðum hætti samhendingar eða liðhendingar. Þat er it sjaunda at hafa eitt málsorð í báðum vísuhelmingum, ok þykkir þat spilla í einstaka vísum. Átta er þat at nýta þótt samkvætt verði við þat er áðr er ort vísuorð eða skemra. Níunda er þat at reka til hinnar fimtu kenningar, er ór ættum er ef lengra er rekit; en þótt þat finnisk í fornskálda verka, þá látum vér þat nú ónýtt. Tíunda er þat at vísu fylgir drag eða stuðill.† Ellipta er þat at 'er' eða 'en' eða 'at' má hafa optar en eitt sinn í vísuhelmingi, svá sem Refr kvað:

Sæll er hinn er hranna
hádýra vel stýrir
(tíð erumk vitnis váða
*víngerð) unir sínu.

Ok svá þó at þat sé í síðara helmingi, ef maðr er nefndr eða kent nafn hans í fyrra helmingi, þótt þat sé eigi nafn annan veg en 'hann' eða 'hinn' eða 'sá' eða 'sjá'. Tólpta er atriðsklauf.

Hvat er tíðaskipti?

Þrent.

Hvernig?

Þat er var, þat er ⟨er⟩, þat er verðr.

Hver setning er þat at breyta háttum með máli einu?

Þat má svá gera at gefa nafn háttum ok greina svá tǫlu háttanna ina fyrstu, en halda annarri ok inni þriðju tǫlu setningar. Þat er sem fyrr var ritat, at hafa átta vísuorð í ørindi, ok hin þriðja tala at hafa sex samstǫfur í vísuorði ok sǫmu setning hendinganna. Háttum er skipt með ýmissum orðtǫkum, ok er þessi einn háttr er kallaðr er sextánmæltr:

Vex iðn. Vellir roðna. **9**
Verpr lind. Þrimu snerpir.
Fæsk gagn. Fylkir eignask.
Falr hitnar. Seðsk vitnir.
*Skekr rǫnd. Skildir bendask.
Skelfr askr. Griðum raskar.
Brandr gellr. Brynjur sundrask.
Braka spjǫr. Litask ǫrvar.

Hér eru tvau mál fullkomin í hverju vísuorði, en orðalengð ok samstǫfur ok hendingar ok stafaśkipti sem dróttkvætt.

Nú er breytt annan veg dróttkvæðum hætti ok enn með máli einu. Þenna hátt kalla menn áttmælt:

Jǫrð verr siklingr sverðum. **10**
Sundr rjúfa spjǫr undir.
Lind *skerr í styr steinda.
Støkkr hauss af bol lausum.
Falla fólk á velli.
Fremr mildr jǫfurr hildi.
Egg bítr á lim lýti.
Liggr skǫr sniðin hjǫrvi.

Hér er mál fyllt í hverju vísuorði. Þessi er hinn þriði:

Ýskelfir kann úlfum **11**
auðmildr búa gildi.
Lætr gylðis kyn gáti
gunnsnarr una harri.
Fær gotna vinr vitni
valbjór afar stóran.

Vargr tér ór ben bergja
blóðdrykk ok grǫn rjóða.

Hér lýkr máli í tveim vísuorðum. Sjá háttr er nú skal rita er hinn fjórði þeira er breyttir eru, en hinn fimti at háttatali; þetta er stælt kallat:

12 Hákun veldr ok hǫldum
(harðráðum guð jarðar
tiggja lér með tíri)
teitr þjóðkonungs heiti.
Vald á víðrar foldar
(vindræfurs jǫfurr gæfu
ǫðlingi skóp ungum)
ǫrlyndr skati gjǫrla.

⟨Hér er svá:⟩

Hákun veldr ok hǫldum
†teitr þjóðkonungs heiti,

en annat ok it þriðja vísuorð er sér um mál, ok er þat stál kallat. Þessi er hinn fimti:

13 Manndýrðir fá mærðar,
mæt ǫld, fira gæti,
lýtr audgjafa ítrum
ǫll. Stóð sær of fjǫllum.
Rjóðvendils gat randa
rœki-Njǫrð at sœkja
(hæf ferð var sú harða)
heim. Skaut jǫrð ór geima.

Þetta kǫllum vér hjástælt. Hér er it fyrsta ⟨vísuorð⟩ ok annat ok þriðja sér um mál, ok hefir þó þat mál eina samstǫfun með fullu orði af *hinu fjórða vísuorði, en þær fimm samstǫfur *er eptir *fara lúka heilu máli, ok skal orðtak vera forn minni.

Þessi er hinn sjaundi:

14 Hákun ræðr með heiðan—
hefir drengja vinr fengit
(lǫnd verr buðlungr brandi
breiðfeld) mikit veldi;

rógleiks náir ríki
remmi-Týr at stýra
(ǫld fagnar því) eignu
—orðróm konungdómi.

Hér hefr upp mál í inu fyrsta vísuorði ok lýkr ⟨í⟩ inu síðarsta, ok eru þau sér um mál. Þessi er inn átti háttrinn:

Þeim er, grundar grímu **15**
gjaldseiðs ok var faldinn,
(drótt man enn *þann) átti
áðr hans faðir ráða.
*Gunnhættir kná grýttu
(gramr býr of þrek) stýra
(stórt ræðr hann, en hjarta
hvetr) buðlunga setri.

Hér er hinn fyrri vísuhelmingr leiddr af þeiri vísu er áðr var kveðin ok fylgir þat málsorð er afleiðing er kǫlluð, er síðarst var í hinni fyrri vísu, þessum vísuhelmi⟨n⟩gi, ok er sá vísuhelmingr eigi elligar rétt⟨r⟩ at máli.

Þess er hinn níundi háttr:

Setr of vísa vitran **16**
vígdrótt—en þar hníga—
(ýr dregsk) við skotskúrum
skjaldborg—í gras *aldir;
vápnrjóðr stikar víða
(vellbrjótr á lǫg) spjótum
(*þryngr at sverða sǫngvi)
sóknharðr þrǫmu jarðar.

Þat málsorð er fyrst er í þessi vísu er síðars⟨t⟩ *í hinni fyrri, ok er hin *síðari svá dregin af hinni fyrri. Því heita þat drǫgur.

Þessi er hinn tíundi háttr er vér kǫllum refhvǫrf. Í þeima hætti skal velja saman þau orðtǫk er ólíkust sé at greina ok hafi þó einnar tíðar fall bæði orð ef vel skal ⟨vera⟩. En til þessa háttar er vant at finna ǫll orð gagnstaðlig, ok er hér fyrir því sum orð dregin til hœginda. *En sýnt er í þessi vísu þat er orðin munu finnask ef vandliga er leitat, ok mun hér þat sýnask at flest frumsmíð stendr til bóta. Svá er hér kveðit:

17 Síks glóðar verr sœkir
slétt skarð hafi jarðar;
hlífgranda rekr hendir
heit kǫld loga ǫldu;
fljótt válkat skilr fylkir
friðlæ—rǫðul⟨s⟩ sævar
ránsið ræsir stǫðvar—
reiðr—glaðr frǫmum meiðum.

Hér er í fyrsta vísuorði svá kveðit: 'síks' ok 'glóðar', 'verr sœkir'. Sík er vatn, glóð er eldr, en eldr ok vatn hatar hvárt ǫðru. 'Verr sœkir': þat er ólíkt at verja ok sœkja. [Ann]at vísuorð er svá: 'slétt skarð hafi jarða⟨r⟩'. Slétt, þat er jafnt, skarð, þa⟨t⟩ er óslétt, ok svá: 'hafi jarðar'. Sær er haf, land er jǫrð. En þá er í *eitt *fall mælt at sá ferr af hafi til jarðar. *Þriðja vísuorð er svá: 'hlífgranda'. Þat er ljóst refhvǫrfmælt, ok svá: 'rekr hendir'. Sá flytr braut er rek⟨r⟩, en sá stǫðvar er hendir. ⟨Svá er it fjórða:⟩ 'heit kǫld', þat er ljós orð, ok svá: 'loga ǫldu'. Logi er eldr, alda er sjár. Fimta orð er svá: 'fljótt válkat'. Fljótt er þat er skjótt er, válkat þat er seint er, ok svá: 'skilr fylkir'. Sá er skilr dreifir, en sá er fylkir samnar. Sétta orð er svá: 'friðlæ'. Friðr er sætt, læ, þat er vél, ok enn: 'rǫðull sævar'. Rǫðull er sól ok gengr hon fyrir eld í ǫllum kenni⟨n⟩gum. Sær er enn sem fyrr í móti eldi. Sjaunda orð er svá: 'ránsið'. Rán, þat er ósiðr, ok svá: 'ræsir stǫðvar'. Sá flytr er ræsir, en sá heldr aptr er stǫðvar. Átta orð er svá: 'reiðr glaðr'. Þat er ljóst mælt, ok svá: 'frǫmum meiðum'. Þat er ójafn⟨t⟩ at vinna manni frama eða meizlur. Hér eru sýnd í þessi vísu sextán orðtǫk sundrgreinilig, ok eru flest ofljós til rétts máls at fœra, ok skal þá svá upp taka: síks glóð, þat er gull; sœkir gulls, þat er maðr; hann verr skarð jarðar hafi slétt, þat eru Firðir, svá heitir fylki í Nóregi; hlífgrandi, þat er vápn; hendir loga ǫldu er maðr, er rekr kǫld heit sverðinu, þat er at hegna ósiðu; fljótt válkat má þat kalla er skjótt ráðit er, þat skilr hann af ófriðinum; konungr heitir fylkir; ránsið ræsir stǫðvar sævar rǫðuls frǫmum meiðum. Þetta heita in mestu refhvǫrf.

Þessi eru ǫnnur refhvǫrf, ok eru hér hálfu færi vísuorð þau er refhvǫrfum eru ort, ok eru þau tvenn í ǫðru vísuorði, ok eru fyrir því kǫlluð in mestu:

18 Blóð fremr (hlǫkk at háðisk
heldr slitnar dul) vitni;
skjǫldr, en skatnar foldir,

skelfr harðr, taka varða;
fal *lætr ⟨of⟩ her hvítan
hollr gramr rekinn framðan,
en tiggja sonr (seggjum
svalr brandr) dugir (grandar).

Hér eru þau refhvǫrf í ǫðru hverju orði: 'heldr' ok 'slitnar', ok 'dul' ok 'vitni'. Dul er laun, en vitni er sannan. En í fjórða vísuorði eru þessi: 'skelfr harðr', 'taka varða'. Í sétta vísuorði er svá: 'hollr gramr', 'rekinn framðan'. Í átta vísuorði er svá: 'svalr brandr' (brandr er elds heiti), 'dugir grandar'. Þetta er ofljóst ort. Hér eru ok ǫnnur máltǫk þau er til máls skal taka, svá at kalla blóð fremr vitni (þat er vargr), en dul eða laun slitnar eða rofnar, at hlǫkk háðisk (þat er orrosta). Ok í ǫðrum fjórðungi er svá, at harðr skjǫldr ⟨skelfr⟩, en skatnar taka varða ríki. Ok ⟨í⟩ þriðja fjórðungi er svá, at hollr gramr of her lætr framðan fal hvítan rekinn; sá er framiðr ⟨er⟩ framar er settr. Í fjórða fjórðungi er svá, at svalr brandr grandar seggjum, en tiggja sonr dugir.

Þessi er hinn þriði refhvarfaháttr:

Segl skekr of hlyn—Huglar— **19**
(hvast drífa skip) rasta,
en fǫll of gram Gylli
grunn (djúp) hata [unna];
ne Rán við⟨r⟩ hafhreinum
háraust—skapar flaustum—
(hrǫnn fyrir húfi þunnum
heil klofnar) frið—deilu.

Þessi eru at kalli in mestu refhvǫrf. Hér er eitt vísuorð í hvárum helmingi þat er refhvǫrfum er ort ok tvenn ór hvárum, svá sem hér: 'grunn djúp', 'hata unna'. ⟨En í⟩ inum efra helmingi er svá: 'heil klofnar', 'frið deilu'. Þessi eru at kalli in mestu refhvǫrf ok minzt af þessum.

Nú hefjask in minni refhvǫrf. Hér eru ein refhvǫrf í vísuorði;

Hélir hlýr at stáli, **20**
hafit fellr, en svífr þelli
(ferð dvǫl firrisk) harða
fram mót lagar glammi;
vindr réttr váðir bendir,
vefr rekr á haf snekkjur,

veðr þyrr; vísa iðjur
(varar fýsir skip) lýsa.

Hér er eitt refhvarf í hverju vísuorði ok flest ofljós. Þessi eru ǫnnur in minni:

21 Lung frá ek lýða þengils
(lá reis of *skut) geisa,
en svǫrð of her herða;
hljóp stóð und gram *Róða.
Þjóð fær þungra skeiða
þrǫng rúm skipat lǫngum;
stál lætr styrjar deilir
stinn kløkk í mar søkkva.

Hér er refhvǫr⟨f⟩ í ǫðru hverju vísuorði. Þessi eru in þriðju:

22 Himinglæva strýkr hávar
(hrǫnn skilja sog) þiljur;
lǫgstíga vill lœgir
ljótr fagrdrasil brjóta;
lýsheims náir ljóma
(líðr ár) of gram blíðum
(uðr rekkir kjǫl *kløkkvan
kǫld) eisa; far geisar.

Hér eru refhvǫrf í hvárum helmingi. Þessi eru in minztu refhvǫrf.

Enn er sá háttr er vér kǫllum refhvarfa bróður:

23 Firrisk hǫnd með harra
hlumr; líðr vetr af sumri,
en flaust við lǫg Lista
lǫng taka hvíld at gǫngu.
Ǫl mœðir lið lýða
(létt skipask hǫll) *it rétta,
en skál at gjǫf góla
gulls svífr (tóm) in fulla.

Hér er í ǫðru ok fjórða þau orð er gagnstaðlig eru sem refhvǫrf, enda standa eigi saman ok er ein samstafa millum þeira ok lúkask bæði eigi ⟨í⟩ eina tíð. Þessir hættir er nú eru ritaðir eru dróttkvæðir at hendingum ok orðalengð: hér eru sex samstǫfur í hverju vísuorði ok aðalhendingar í ǫðru ok inu fjórða en skothendur í fyrsta ok þriðja.

Hvernig skal skipta dróttkvæðum hætti með hendi⟨n⟩gum eða orðalengð?

Svá sem hér er:

Hreintjǫrnum gleðr horna 24
(horn ná lítt at þorna,
mjǫðr hegnir bǫl bragna)
bragningr skipa sagnir;
fólkhǫmlu gefr framla
framlyndr viðum gamlar,
hinn er heldr fyrir skot skjǫldum,
skjǫldungr hunangs ǫldur.

Hér er þat málsorð fyrst í ǫðru ok inu fjórða vísuorði er síðarst er í inu fyrsta ok þriðja. Þetta er tilsagt:

Rǫst gefr ǫðlingr jastar 25
—ǫl virði ek svá—fyrðum.
Þǫgn fellir brim bragna
—bjórr forn er þat—horna.
Máls kann mildingr heilsu
—mjǫðr heitir svá—veita.
Strúgs kemr í val veiga
—vín kallak þat—galli.

Nú er orðskviðuháttr:

Fúss brýtr fylkir eisu 26
fens—bregðr hǫnd á venju.
Ránhegnir gefr Rínar
rǫf—spyrr ætt at jǫfrum.
Mjǫk trúir ræsir rekka
raun—sér gjǫf til launa.
Ráð á lofðungr lýða
lengr—vex hverr af gengi.

Ískalda *skar ek ǫldu 27
eik (var súð in bleika
reynd) til ræsi[s fundar]
ríks. Emk kuðr at slíku.
Brjótr þá hersirs heiti
hátt (dugir sœmð at vátta)

auðs af jarla prýði
ítrs. Vara siglt til lítils.

Þetta er álagsháttr. Hér hefr upp annat ok hit fjórða vísuorð með fullu orði ok einni samstǫfu, ok leiðir þat orð af hinu fyrra vísuorði, en þær fimm samstǫfur er þá eru eptir eru sér um mál.

Þessi er hinn fyrsti háttr er ritaðr sé þeira er breytt er af dróttkvæðum hætti með fullu háttaskipti, ok heðan frá skal nú rita þær greinir er skipt er dróttkvæðum hætti ok breytt með hljóðum ok hendingaskipti eða orðalengð, stundum við lagt en stundum af tekit. Þetta er tvískelft:

28 Vandbaugs veitti sendir
vígrakkr—en gjǫf þakkak
skjaldbraks skylja mildum—
*skipreiðu mér—heiða;
fann næst fylkir unna
fǫl dýr at gjǫf stýri
stálhreins; styrjar deilis
stórlæti sá ek mæta.

Hér ⟨er⟩ í fyrsta ok þriðja vísuorði þat er háttum skiptir: hér standask hljóðfyllendr svá nær at ein samstafa er í milli þeira. Þeir gera skjálfhendur stuðlar ok er hinn fyrri upphaf vísuorðs, en hendingar standask sem first. En ef frumhending er í þeiri samstǫfu er næst er hinni fyrstu, þá bregzk eigi skjálfhenda.

Þessi er detthendr háttr:

29 Tvær man ek hilmi hýrum
heimsvistir ótvistar,
hlaut ek ásamt at sitja
seimgildi fémildum;
fúss gaf fylkir hnossir
fleinstýri margdýrar,
hollr var hersa stilli
hoddspennir fjǫlmennum.

Hér skiptask hættir í ǫðru ok fjórða vísuorði, ok ræðr in fjórða samstǫfun háttunum. Þetta er draugsháttr:

30 Þoll bið ek hilmis hylli
halda grœnna skjalda,

askr beið af því þroska
þilju Hrungnis ilja;
vígfoldar njót valdi
vandar margra landa—
nýtr vartu oss—til ítrar
elli dólga fellir.

Hér er enn í fjórða ok í ǫðru vísuorði þat er háttum skiptir, ok ræðr hér hin þriðja samstafa.

Nú hefr upp annat kvæði:

Stáls dynblakka støkkvi 31
stinngeðs samir minnask
(álms bifsœki aukum
Yggs *feng) á lof þengils;
odds bláferla jarli
ǫrbrjót ne skal þrjóta
(Hárs saltunnu hrannir
hrœrum) óð at stœra.

Þetta heitir bragarbót. Hér skiptir háttum í fyrsta ok þriðja vísuorði. Hér standask sem first má stuðlar, en hendingar svá at ein samstafa er á milli. Þat greinir háttuna.

Þenna hátt kalla menn riðhendur:

Él þreifsk skarpt of Skúla 32
skýs snarvinda lindar,
egg varð hvǫss í hǫggum
hræs dynbrunnum runnin;
sveimþreytir bjó sveita
snjallr ilstafna hrafni;
Páll varð und fet falla
fram þrábarni arnar.

Hér skiptir háttum í ǫðru ok fjórða vísuorði: standa þar hendingar báðar samt nær enda ok lúkask á einu[m hljóðsta]f báðar, ok er betr at samhljóðandi *sé eptir aðra.

Þessi háttr er kallat veggjat:

Lífs varð rán at raunum 33
(réð sverð) skapat mjǫk ferðum,
stǫng óð þrátt á þingi
þjóðsterk, liðu fram merki;

hrauð of hilmis bróður
hvǫss egg friðar ván seggjum,
spjót náðu blá bíta,
búandmenn hlutu þar renna.

Hér er háttaskipti í ǫðru ok fjórða vísuorði ok er þar ein samstǫfun sett í svá at tvær eru síðar ok aukit því lengð orðsins. Nú er flagðaháttr:

34 Flaust bjó fólka treystir
fagrskjǫlduðustum ǫldum,
leið skar bragnings bróðir
bjartveggjuðustu reggi;
hest rak hilmir rasta
harðsveipaðastan reipum,
sjár hlaut við þrǫm þjóta
þunghúfuðustu lungi.

Hér skiptir háttum í ǫðru ok inu fjórða vísuorði: er hér aukit bæði samstǫfu ok fullnat orðtak sem framast, ok eptir þá samstǫfun eru þrjár samstǫfur ok er rétt dróttkvætt ef hon er ór tekin.

Þessi háttr er in forna skjálfhenda:

35 Reist at Vágsbrú vestan
(varrsíma bar fjarri)
heitfastr hávar rastir
hjálm-Týr svǫlu stýri;
støkr óx er bar blakka
brims fyrir jǫrð it grimma
herfjǫlð (húfar svǫlðu)
*hrannláð búandmanna.

Hér er skjálfhent með aðalhending í þriðja vísuorði í hvárum tveggja helmingi, en ⟨at⟩ ǫðru sem dróttkvætt. Þenna hátt fann fyrst Veili. Þá lá hann í útskeri nokkvoru, kominn af skipsbroti, ok hǫfðu þeir illt til klæða ok veðr kalt. Þá orti hann kvæði er kallat er kviðan skjálfhenda eða drápan steflausa, ok kveðit eptir Sigurðar sǫgu.

Þetta er þríhent kallat:

36 Hristi hvatt þá er reistisk
herfǫng mjǫk lǫng véstǫng,
samði fólk en frǫmðusk
fullsterk hringserk grams verk;

hǫnd lék—*herjum reyndisk—
hjǫrr kaldr—allvaldr mannbaldr;
egg frá ek breiða bjoggju
bragning fylking; stóð þing.

Þrennar aðalhendingar eru hér í ǫðru ok inu fjórða vísuorði, ok lúkask allar einnig, ok fylgir samstǫfun fyrir hverja. Nú er hinn dýri háttr:

Vann (kann virðum banna 37
vald) gjald (hǫfundr aldar)
ferð verð fólka herði
fest mest (sá er bil lestir);
hátt þrátt hǫlða áttar
—hrauð auð jǫfurr rauðum—
(þat) gat þengill skatna
þjóð (stóð af gram) bjóða.

Hér eru í fyrsta ok þriðja vísuorði tvær aðalhendingar samt í upphafi, en hin þriðja at hætti við enda.

*Farar snarar fylkir byrjar, **38†**
freka breka lemr á snekkjum,
vaka taka vísa ⟨rekkar,
viðar⟩ skriðar at þat biðja;
svipa skipa sýjur heppnar
sǫmum þrǫmum í byr rǫmum,
Haka skaka hrannir blǫkkum
hliðar, miðar und kjǫl niðri.

⟨Hér eru þrjár hendingar í vísuorði ok skothend í fyrsta ok þriðja vísuorði in þriðja hending ok fylgir samstafa hverri hendingu.⟩

Þessi háttr er kallat tiltekit:

Ok hjaldreifan hófu **39**
hoddstiklanda miklir
(morðflýtir kná mœta
málmskúrar dyn) hjálmar
hjaldrs þá er hilmir foldar
hugdýrum gaf stýri
(ógnsvellir fær allan)
jarldóm (gǫfugr sóma).

Hér skiptir háttum it fimta vísuorð ok leiðir í því orði máltak af fyrra vísuhelmingi ok dregsk þat vísuorð með hljóðfylling mjǫk eptir skjálfhendu inni ný[ju].

Þessi háttr er kallat greppaminni:

40 Hverr fremr hildi barra?
Hverr er mælingum ferri?
Hverr gerir hǫpp at stœrri?
Hverr kann auð at þverra?
Veldr hertogi hjaldri,
hann er first *blikurmanni,
hann á hǫpp at sýnni,
hann vélir blik spannar.

*Þessum *hætti er breytt til dróttkvæðs háttar með orðum. Nú er sá háttr er vér kǫllum liðhendur:

41 Velr ítrhugaðr ýtum
otrgjǫld jǫfurr snotrum,
opt hefr þings fyrir þrøngvi
þungfarmr Grana sprungit;
hjǫrs vill rjóðr at ríði
reiðmálmr Gnitaheiðar,
vígs er hreytt at hættis
hvatt Niflunga skatti.

Þat eru liðhendur er hinn sami stafr stendr fyrir hendingar, ok þá er rétt ort liðhendr háttr at í ǫðru ok í hinu fjórða vísuorði sé oddhending ok skothending við þær hendingar er í *inu ⟨fyrra⟩ vísuorði eru, ok verðr þá einn upphafsstafr allra þeira þriggja hendinganna.

Nú er sá háttr er vér kǫllum rétthent:

42 Alrauðum drífr auði,
ógnrakkr firum hlakkar,
veit ek, *hvar vals á reitu
verpr hringdropa snerpir;
snjallr lætr á fit falla
fagrregn jǫfurr þegnum
(ógnflýtir verr ýtum
arm) Mardallar hvarma.

Hér eru aðalhendingar í fyrsta ok þriðja vísuorði, en gætt at taka ór skothendur.

Enn er sá háttr er vér kǫllum ina minni *alhendu, þat eru skothendur í inu fyrsta vísuorði í báðum helmingum, svá sem hér segir:

Samþykkjar fremr søkku⟨m⟩ 43
snarr Baldr hjarar aldir,
gunnhættir kann Grotta
glaðdript hraða skipta;
féstríðir kná Fróða
friðbygg liði tryggva,
fjǫlvinjat hylr Fenju
falr meldr alinveldi.

In minni alhenda er þá rétt ort at haldit sé vísulengð saman, en ef henni er skotit í fulla alhendu svá at skothendur sé þar sumar eða allar í vísuorði, þá er þat eigi rétt.

Nú er alhent:

Frama skotnar gram; gotnum 44
(gjǫf sannask) rǫf spannar—
menstiklir—vensk mikla—
manndýrðir vann skýrðar;
herfjǫlð—bera hǫlðar—
hagbáls lagar stála
friðask sjaldan við valdi—
vallands svala branda.

Hér eru tvennar aðalhendingar í hverju vísuorði. Þessi þykkir vera fegrstr ok vandastr, ef vel er ortr, þeira hátta er kvæði eru ort eptir, ok er þá full alhending ef eigi finn⟨sk⟩ í 'at', 'ek', 'en', eða þau smáorð er þeim fylgja, nema þau standi í hendingum, en eigi hafa allir menn þat varazk, ok er þat fyrir því eigi rangt, sem kvað Klœingr byskup:

Bað ek sveit á glað Geitis,
gǫr er íð at fǫr tíðum,
drǫgum hest á lǫg lesta,
lið flýtr en skrið nýtum.

Þetta er stamhendr háttr:

Lætr undin brot brotna 45
bragningr fyrir sér hringa,

sá tekr fyrir men menja
mætt orð of sik fættir;
armr kná við *blik blikna
brimlands viðum randa
þar er hǫnd at lið liðnar
lýslóðar berr glóðir.

[Hér er í] fyrsta ok þ[riðja vísuorði tv]íkveðit at einni samstǫfu ok haft þat til hendinga, ok fyrir því kǫllum vér þetta stamhent at tvíklypt er til hendingarinnar, ok standa svá hendingar í orðinu sem riðhendur.

Nú er sá háttr er samhent er kallat:

46 Virðandi gefr virðum
*verbál liðar skerja,
gleðr vellbrjóti vellum
verðung *afar þungum;
ýtandi fremr ýta
auðs sæfuna rauðum
þar er mætum gram mæti
marblakks skipendr þakka.

Hér eru þær hendingar er vér kǫllum samhendur, því at þessar eru allar með einum stǫfum ok eru í *fyrsta ok þriðja vísuorði svá settar sem skothendur í dróttkvæðum hætti.

Nú er iðurmælt:

47 Seimþverrir gefr seima
seimǫrr liði beima,
hringmildan spyr ek hringum
hringskemmi brott þinga;
baugstøkkvir fremr baugum
bauggrimmr hjarar draug⟨a⟩,
viðr gullbroti gulli
gullhættr skaða fullan.

Hér er þrim sinnum haft *samhending, tysvar í fyrsta ok þriðja vísuorði, en í ǫðru ok hinu fjórða er haldit afhending sem í dunhendum hætti.

Þessi háttr heitir klifat:

48 *Auðkendar verr auði
auð-Týr boga nauðir,

*þar er auðviðum auðit
auðs í gulli rauðu;
heiðmǫnnum býr heiðis
heiðmildr jǫfurr reiðir,
*venr heiðfrǫmuðr heiðar
heiðgjǫf vala leiðar.

Hér halda samhendingar of allan vísuhelming ok taka með aðalhending ina síðari í ǫðru ok inu fjórða vísuorði.

Nú eru þeir hættir er stúfar heita:

Hjaldrremmir tekr Hildi 49
(hringr brestr at gjǫf) festa,
hnígr und Hǫgna meyjar
hers valdandi tjald;
Heðins mála býr hvílu
hjálmlestanda flestum,
morðaukinn þiggr mæki
mund Hjaðninga sprund.

Hér er it fjórða vísuorð stýft ok tekin af samstafa er í dróttkvæðum hætti skal setja með hending. Þessi er meiri stúfr:

Yggs drósar rýfr eisa 50
ǫld móðsefa tjǫld,
glóð støkkr í hof Hlakkar
hugtúns firum brún;
geðveggjar svífr glugga
glæs dynbrími hræs,
hvattr er hyrr at slétta
hjaldrs gnapturna aldrs.

Hér er stýft annat ok it fjórða vísuorð. Nú er hinn mesti stúfr; hér eru ǫll vísuorð stýfð:

Herstefnir lætr hrafn 51
hungrs fullseðjask ungr,
ilspornat getr ǫrn
aldrlausastan haus;
vilja borg en vargr
vígsára klífr grár,
opt sólgit fær ylgr
(jǫfurr góðr vill [sv]á) blóð.

Þessir hættir er nú eru ritaðir *eru greindir í þrjá staði, þvíat menn hafa ort fyrr svá at í einni vísu var annarr helmingr stýf⟨ð⟩r en annarr helmingr tvístýf⟨ð⟩r, ok eru þat háttafǫll. Sá er hinn þriði er alstýfðr er, því at hér eru ⟨ǫll⟩ vísuorð stýfð.

Nú skal rita þann hátt er skothendr heitir:

52 Sær skjǫldungs niðr skúrum,
skǫpt darraðar lyptask,
hrindr gunnfana grundar
glygg *of frœk[num tiggja];
geisa [vé fyr vísa,
veðr stǫng] at hlym Gungnis,
styrk eru mót und merkjum
málms of ítran hilmi.

Hér eru skothendur í ǫllum vísuorðum, en annat sem dróttkvæðr háttr. Nú er sá háttr er vér kǫllum liðhendur:

53 Stjóri vensk at stœra
stór verk dunu geira,
halda kann með hildi
hjaldr-Týr und sik foldu;
harri slítr í hverri
Hjarranda fǫt snerru,
falla þar til fyllar
fjallvargs jǫru þollar.

Í þessum hætti eru liðhendur með tvennu móti, en aðrar á þá lund at við ina fyrri hending í fyrsta ok þriðja vísuorði ⟨. . .⟩

Nú skal rita þá háttu er fornskáld hafa kveðit ok eru nú settir saman, þótt þeir hafi ort sumt með háttafǫllum, ok eru þessir hættir dróttkvæðir kallaðir í fornum kvæðum, en sumir finnask í lausum vísum, svá sem orti Ragnarr konungr loðbrók með þessum hætti:

54 Skýtr at Skǫglar veðri
(en skjaldagi haldask)
Hildar hlemmidrífu
of hvítum þrǫm rítar;
en í sœfis sveita
at sverðtogi ferðar
rýðr aldar vinr odda
(þat er jarlmegin) snarla.

Hér ⟨er⟩ í fyrsta ok þriðj⟨a⟩ vísuorði háttlausa, en í ǫðru ok fjórða aðalhendingar, en hǫfuðstafrinn stendr svá, sá er kveðandi ræðr, í ǫðru ok inu fjórða vísuorði, at þar er fyrir sett samstafa ein eða tvær, en at ǫðru sem dróttkvætt.

Nú skal rita Torf-Einars hátt:

Hverr séi jǫfra œgi **55**
jarl fjǫlvitrum betra,
eða gjarnara at gœða
glym hraðsveldan skjalda?
Stendr *af stála skúrar
styrr ólítill Gauti
þá er fólks *jaðarr foldir
ferr sig-Njǫrðum varða.

Hér er í fyrsta ok þriðja vísuorði háttlausa, en í ǫðru ok fjórða skothent ok riðhent. Nú er Egils háttr:

Hverr ali blóði byrsta **56**
bens rauðsylgjum ylgi
nema svá at gramr of gildi
gráð dag margan vargi?
Gefr oddviti undir
egg nýbitnar vitni,
herr sér Fenris fitjar
fram klóloðnar roðna.

Hér er í fyrsta ok þriðja vísuorði háttlausa, en í ǫðru ok inu fjórða aðalhendingar ok riðhent. Nú er Fleins háttr:

Hilmir hjálma skúrir **57**
herðir sverði roðnu,
hrjóta hvítir askar,
hrynja brynju spangir;
hnykkja Hlakkar eldar
harð[a] svarðar landi,
remma rimmu glóðir
randa grand of jarli.

Hér er svá farit hendingum sem í dróttkvæðum hætti, en hendingar eru settar saman í ǫndurðu vísuorði. Nú er Braga háttr:

58 Er til hjálma hyrjar
herjum styrjar væni
þar svá at jarl til ógnar
egnir tognu sverði;
sjá *kná garð fyrir grundu
grindar Þundar jaðra
er skatna vinr skjaldar
skyldisk galdr at fremja.

Hér er í fyrsta ok þriðja vísuorði it síðarsta málsorð haft til hendingar, en missir þess orðs ins fyrra er gera skyldi skothending, en við þetta hendingarorð eru í ǫðru ok *inu fjórða vísuorði hendingar, ok er þat ǫnnur hending skothenda ok lið⟨hending en ǫnnur aðalhending við⟩ ina fyrstu, en þessar hendingar er standa í ǫðru ok inu fjórða vísuorði† standa sem í Fleins hætti, [Víða er þat í] fornskálda verka er í einni vísu eru ymsir hættir eða háttafǫll, ok má eigi yrkja eptir [því] þó at þat þykki eigi spilla í fornkvæðum.

Nú eru þeir hættir, greindir í þrjá staði, er kimblabǫnd heita. Þessi er einn:

59 Hjálmlestir skerr Hristar
hreggǫld Sigars veggi,
gramr lætr í byr brjóta
brands hnigþili randa stranda;
stálhrafna lætr stefnir
styrvind of sik þyrja,
þiggr at Gǫndlar glyggvi
*gagn oddviti bragna sagna.

Hér er í fjórða vísuorði í hvárum helmingi aukit aðalhending með tveim samstǫfum eptir vísuorð, en at ǫðru sem dróttkvætt. Nú er it meira kimblaband:

60 Álmdrósar skylr ísa
ár flest meginbára sára,
kœnn lætr hræ[s] á hrǫnnum
hjálmsvell jǫfurr gella fella;
styrjǫkla kná stiklir,
stinn, mens legi venja benja,
lætr stillir frár fylla
fólk sund hjarar lunda unda.

Hér eru tvenn kim⟨b⟩labǫnd á hvárum helmingi. Þessi eru in mestu kim⟨b⟩labǫnd:

Hræljóma fellr hrími (tími 61
hár vex of gram sára ára)
—frost nemr—of *hlyn Hristar—Mistar
herkaldan þrǫm skjaldar aldar;
gullsendir brýtr grundar Hrundar
gunnveggs stǫfum leggi hreggi,
sóknvallar (spyr ek) svelli (elli)
svá skotnar þat (gotna þrotna).

Hér fylgir hverju vísuorði kimblaband.

Nú skal rita hrynjandi háttu; þessi er hinn fyrsti:

Tiggi snýr á ógnar áru 62
(undgagl veit þat) sóknar hagli,
yngvi drífr at hreggi hlífa
(hjǫrr vélir fjǫr) brynju éli;
vísi heldr of fjǫrnis foldir
(fólk skipta svá) boga driptum,
skúrum lýstr of hilmi hraustan
(hans fregnum styr) Mistar regni.

Hér er it fyrsta ok þriðja vísuorð aukit framan tveim samstǫfum til háttar setningar, en ef þær eru af teknar þá er eptir sem dróttkvætt, en ór ǫðru ok fjórða vísuorði má taka málsorð þat er tvær samstǫfur fylgja, in fimta ok in sétta í vísuorði; þá er þat orð ok dróttkvætt. Í hrynhendum háttum eru optast átta samstǫfur í vísu⟨orði⟩, en hendingar ok stafaskipti fara sem í dróttkvæðum hætti. Þetta kǫllum vér dróttkvæða hrynjandi.

Nú skal sýna fleiri skipun háttanna. Er þessi hrynhenda kǫlluð trollsháttr:

Stála kend⟨i⟩ steykkvilundum 63
styrjar valdi rauðu falda,
rekkar stýrðu rétt til jarðar
roðnu barði austan fjarðar;
oddum rendi eljunstrandir
ýta ferðar hringa skerðir,
hilmir stœrði hvǫssu sverði
heila grundar meginundir.

Hér eru átta samstǫfur ⟨í hverju vísuorði; hér eru hluthendur í ǫllum orðum ok fylgja *þrjár samstǫfur⟩ hverri hendingu, ok svá fara skothendur ok aðalhendingar ok stafaskipti sem í hrynhendu.

Þessi er einn hrynhendr háttr:

64 Vafði lítt er virðum mœtti
vígrœkjandi fram at sœkja,
sk[erðir gekk í s]kúrum Hlakk[ar
Skǫglar serks fyrir] roðnum merkjum;
ruddisk land en ræsir Þr[œnda]
Ribbungum skóp bana [þun]gan,
Gunnarr skaut und Gera fótar
grimmsett[a il hjar]na kletti.

Þetta er hrynhenda óbr[eytt]. Þetta er draughent:

65 Vápna hríð velta náði
vægðarlaus feigum hausi,
hilmir lét hǫggum mœta
herða klett bana verð⟨a⟩n;
fleina lands fylkir rendi
fjǫrnis hlíð meginskíði
(*ǫflugt *sverð *eyddi fyrðum
jǫfri kent) holdi fenta.

⟨Í⟩ þessum hætti eru tíðast sjau samstǫfur ⟨í⟩ hverju vísuorði, en hendingar ok stafaskipti sem í dróttkvæðum ⟨hætti⟩, ok ef hér er ór tekin ein samstǫfun fyrsta eða þriðja vísuorði sú er stendr næst hinni fyrstu, þá falla hljóðin ǫll sem í dróttkvæðum hætti. Svá má ok af taka í ǫðru ok hinu fjórða vísuorði ina sǫmu samstǫfun ok er þá þat dróttkvætt; ok verðr sumt eigi mjúkt.

Þenna hátt kǫllum vér munnvǫrp:

66 Eyddi úthlaupsmǫnnum
ítr hertogi spjótum,
sungu stál of stillis
(stóð ylgr í val) dólgum;
hal margan lét hǫfði
hoddgrimmr jǫfurr skemra,
svá kann rán at refsa
reiðr oddviti þjóðum.

Hér er háttlausa ⟨í⟩ inu *fyrsta ok þriðja vísuorði, en í ǫðru ok inu fjórða skothendur. Nú er sá háttr er kallaðr er háttlausa:

Ortak ǫld at minnum 67
þá er alframast vissak
of siklinga snjalla
með sex tøgum hátta.
Sízt hafa veg né vellum
er virðan mik létu
á aldinn mar orpit
(þat er oss frami) jǫfrar.

Í þessum hætti eru øngvar hendingar, en stafaskipti sem í dróttkvæðum hætti. Nú eru saman settir í tveim kvæðum sex tigir hátta ok um fram þær átta greinir er fyrst er skipat ⟨í⟩ dróttkvæðum hætti með málsgreinum þeim er fylgja hættinum, ok eru þessir hættir allir vel fallnir til at yrkja kvæði eptir ef vill.

Nú skal upp hefja it þriðja kvæði þat er ort er eptir inum smærum háttum, ok eru þeir hættir þó margir áðr í lofkvæðum. Hér hefr upp tøgdrápulag:

Fremstr varð Skúli . . . 68
Skala lof dvala,
sem ek mildum gram
mærð fjǫlsnœrða;
meir skal ek stœri
styrs hróðr fyrir
(kærr var ek harra)
*hers gnótt bera.

Hér er í ǫðru ok í fjórða vísuorði fjórar samstǫfur ok tvær aðalhendingar ok svá settr hǫfuðstafr sem í dróttkvæðu, en í fyrsta ok þriðja vísuorði eru ok fjórar réttar samstǫfur ok in fimta afkleyfissamstafa, þat er 'ek' eða 'af' eða 'en' eða 'er' eða þvílíkt. Þar eru ok skothendingar ok ein hljóðfylling við hǫfuðstafinn.

Þetta er annat tøglag:

Kunn bjó ek kvæði 69
konungs bróður þjóð
—þann veit ek þengil—
þrenn—fjǫlmennan;

fram skal in fjórða
fólkglaðs vaða
ljóss elds lagar
lofun friðrofa.

Svá ferr hér *annat ok fjórða vísuorð sem í fyrra hætti, en it fyrsta ok þriðja vísuorð er hér hendingalaust, en tveir hljóðfyllendr við hǫfuðstaf sem í dróttkvæðu.

Þessi er hinn þriði háttr, er vér kǫllum [hagmælt]:

70 [Mitt] er of mœti
[mart lag bra]gar
áðr ókveðit
oddbraks spakan;
hlýtr [gram]s geta
greppr óhnepp[ra
ský]rr skrautfara.
. . . skjǫldunga ungr.

Í þessum hætti eru skot[hendingar] í fyrsta ok þriðja vísuorði ok stafaskipti sem í dróttkvæðum hætti, en at ǫðru sem tøgmælt. Í ǫllu tøglagi er eigi rangt þótt fimm samstǫfur sé í vísuorði er skammar eru sumar ok skjótar. Þat er tøgdrápuháttr at stef skal vera til fyrsta vísuorðs ok lúka því máli í inu síðarsta vísuorði kvæðisins, ok er rétt at setja kvæðit með svá mǫrgum stefjamélum sem hann vil, ok er þat tíðast at hafa ǫll jafnlǫng, en hvers stefjaméls skal stef upphaf ok niðrlag.

Nú er grœnlenzki háttr:

71 Slóð kann sneiðir
seima geima
hnigfák Haka
hleypa greypa,
hinn er af hlunni
hesta festa
lætr leyfðr skati
langa ganga.

Hér er hit fyrsta ok þriðja vísuorð svá sem hagmælt, en annat ok fjórða með aðalhendingum, ok eru tvær samstǫfur aðalhendar á ok *endask báðar í einn staf.

Nú er hinn skammi háttr:

Gull *kná—greppar— 72
glóa—róa,
váss eru seggir
samir framir;
eik má und jǫfri
una bruna,
þá nýtr vísi
viðar skriðar.

Hér er it fyrsta ok þriðja vísuorð hendingalaust, en annat ok it fjórða sem grœnlenzki háttr ok skemri orðtǫkin. Nú er nýi háttr:

Ræsir glæsir 73
Røkkva *døkkva
hvítum rítum
hreina reina,
skreytir hreytir
skafna stafna
hringa stinga
hjǫrtum svǫrtum.

Í þessum hætti eru í hverju vísuorði fjórar samstǫfur en tvær aðalhendingar ok lúkask í einn staf báðar ok engi afkleyfisorð. Þetta er stúfhent:

Hafrǫst hristir 74
hlunnvigg tiggja,
borðgrund bendir
brimdýrs stýri;
blá veit brjóta
byrskíð víði
*bǫðharðr *bǫrðum
buðlungr þungan.

Í þessum hætti eru fjórar samstǫfur í vísuorði en hendingar ok stafaskipti sem í dróttkvæðum hætti nema þat at allar hendingar eru náhendar. Þetta er náhent:

Hrinda lætr hniggrund 75
hafbekks snekkjur,
þá er falla, fleinþollr
frár, mál, stálum;
hlumi lítr *hergramr
hirðmenn spenna

—en rœði *raungóð—
rógálfr—skjálfa.

Í þessum hætti eru fjórar samstǫfur í vísuorði ok er eigi rangt ⟨í⟩ inu fyrsta ok þriðja þótt fimm sé; þar eru skothendur; í ǫðru ok inu fjórða eru aðalhendingar ok báðar saman ok in fyrri stýfð, en stafaskipti sem í dróttkvæðu.

Þetta er hnugghent:

76 Hrannir strýkva hlaðinn bekk,
haflauðr skeflir,
kasta náir kjalar stíg
kalt hlýr sǫltum;
svǫrtum hleypir svana fjǫll
snjallmæltr stillir
hlunna of Haka veg
hríðfeld skíðum.

Hér er í fyrsta ok þriðja vísuorði sjau samstǫfur ok hendingalaust en rétt at stǫfum, en annat ok it fjórða hefir fjórar samstǫfur en rétt at stǫfum ok skothending ok oddhent ok stýfð in fyrri hending.

Nú er hálfhnept:

77 [Snyðja] lætr í sólroð
snekkjur á Manar hlekk,
árla sér ungr jarl
allvaldr [brek]a fall;
lypta[sk kn]á lýðr opt
lauki of kjalar raukn,
greiða náir glygg váð,
greipum mœta dragreip.

Í þessum hætti eru sex samstǫfur í vísuorði, en eigi er rangt þótt verði fimm eða sjau. Í fyrsta ok þriðja vísuorði eru skothendur, en aðalhendingar í ǫðru ok hinu fjórða; í hvárumtveggja stað in fyrri hending rétt í dróttkvæðu en in síðari stýfð eða hnept, þat er allt eitt.

Sjá háttr er alhneptr:

78 Hrǫnn skerr—hvatt ferr—
húfr kaldr—allvaldr,
lá brýtr—lǫg skýtr—
limgarmr—rangbarmr;
brátt skekr—byrr rekr—
blán vegg—rásk egg,

jarl lætr almætr
ósvipt húnskript.

Í þessum hætti eru fjórar samstǫfur í vísuorði ok tvær aðalhendingar ⟨ok lúkask⟩ báðar í einn staf ok allar hendingar hneptar. Þetta er Haðar lag:

Læsir leyfðr vísi **79**
landa útstrandir
blíðr ⟨ok⟩ bláskíðu⟨m⟩
barða randgarði;
ern kná jarl þyrna
oddum valbrodda
jǫrð með élsnœrðum
jaðri hrænaðra.

Í þessum hætti eru fimm samstǫfur í vísuorði en hendingar ok stafaskipti sem í dróttkvæðum hætti.

Nú eru þeir hættir er runhendur eru kallaðar. Þeir eru með einu móti: hverr háttr runhendr skal vera með aðalhendingum tveim ok í sínu vísuorði hvár hending. Þessi er rétt runhenda:

Lof er flutt fjǫrum **80**
fyrir gunnǫrum
(né spurð spǫrum
spjǫll grams) snǫrum;
hefi ek hans fǫrum
hróðrs ǫrum
ypt óvǫrum
fyrir auðs bǫrum.

Þessi háttr er haldinn með einni hending í hverju vísuorði; ok svá er sú runhending er skilr hendingar ok skiptir orðum. Því er þetta runhent kallat.

Þetta er hin minni runhenda:

Fluttak frœði **81**
*of frama grœði
(tunga tœði)
með tǫlu rœði;
stef skal stœra
stilli Mœra

(hróðr dugir hrœra)
ok honum fœra.

Hér gengr hending of há⟨lfa⟩ vísu en ǫnnur í síðari helmingi.

Þessi háttr er stýfðr eða hneptr af inum fyrra; þessi er in minzta runhenda:

82 Slíkt er svá,
siklingr á
(ǫld þess ann)
*orðróm þann;
jarla er
austan ver
skatna skýrstr
Skúli dýrstr.

⟨Í⟩ þessum hætti eru þrjár samstǫfur í vísuorði, en tvau vísuorð sér um hending; stafaskipti sem í dróttkvæðu. Enn finnsk þat svá at eigi er rangt ef stendr einu sinni fyrir málsorð hljóðstafr sá er kveðandi ræðr.

Þessir eru enn runhendir:

83 Naðrs gnapa ógn alla,
eyðir baugvalla,
hlunns of hástalla
hestar svanfjalla;
orms er glatt galla
með gumna spjalla;
jarl fremr sveit snjalla,
slíkt má skǫrung kalla.

Þessi háttr er ort með fullri runhending ok eru þar tíðast fimm samstǫfur í vísuorði, eða sex ef skjótar eru. Þessi er annarr:

84 Orð fekk gott gramr,
hann er gunntamr,
mjǫk er fullframr
fylkir rausnsamr;
hinn er mál metr
milding sízt getr
þann er svá setr
seggi hvern vetr.

Þessi er hneptr [af inni fyrri] runhending.

Mærð vilk auka 85
M[istar] lauka
góma [sverði]
grundar skerði;
dýrð skal segja
(drótt má þegja)
styrjar glóða
støkk[vi-Mó]ða.

Í þeima hætti eru fjórar samstǫfur í hverju vísuorði en hǫfuðstafr sem í dróttkvæðum hætti ok fylgir þeim einn hljóðfyllandi.

Þessi er hinn þriði háttr runhendr:

Veit ek hrings hraða 86
í hǫll laða
(gott ⟨er⟩ hús Hlaða)
hirð ǫlsaða;
*drekkr gramr glaða—
en at gjǫf vaða
vitar valstaða—
vandbaugskaða.

Þetta er rétt runhending, ok er þessi háttr tekinn af tøglagi. Hér eru fjórar samstǫfur í vísuorði eða fimm ef skjótar eru.

Þessi er hinn minni runhenda:

Drífr handa hlekkr 87
þar er hilmir drekkr,
mjǫk er brǫgnum bekkr
blíðskálar þekkr;
leikr hilmis her
hreingullit ker
(segi ek allt sem er)
við orða sker.

Þessi er hneptr af hinum fyrra. Þessi er in minzta:

En þá er hirð til hallar 88
her⟨s⟩ oddviti kallar,
opt tekr jarl at fagna
við ótali bragna;
búin er gjǫf til greizlu
at gullbrota veizlu,

þrǫngt sitr þjóðar sinni,
þar er mestr frami inni.

Þessi runhenda er tekin af dróttkvæðum hætti ok eru hér jafnmargar samstǫfur ok svá stafaskipti sem í dróttkvæðu.

Nú hefr upp inn fjórða bálk runhendinga:

89 Hirð gerir hilmis kátt,
hǫll skipask þrǫngt at gátt,
auð gefr þengill þrátt,
þat spyrr fram í átt;
slíkt tel ek hilmis hátt,
hans er rausn of mátt,
jarl brýtr sundr í smátt
slungit gull við þátt.

Þessi háttr er hneptr af inum fyrra ok rétt runhendr. Nú er minni runhenda:

90 Mǫrg þjóð ferr til siklings sala,
sœmð er þar til allra dvala,
tiggi veitir seim⟨a⟩ svala,
satt er bezt of hann at tala;
bresta spyrjum bauga flata
—bragna vinr kann gulli hata—
(œðri veit ek at gjǫflund gata
grundar vǫrðr) fyrir hringa skata.

Þessi runhenda er tekin af hrynhendum hætti.

91 Þiggja kná með gulli glǫð
gotna ferð at ræsi mjǫð,
drekka lætr hann sveit at sín
silfri skenkt it fagra vín;
greipum mœtir gullin skál,
gumnum sendir Rínar bál
—eigi hittir œðra mann—
jarla beztr—en skjǫldung þann.

Þessi er hneptr af hinni fyrri runhendu.

Hér hefr upp hinn fimta runhendan bálk:

92 Getit var grams fara,
gert hefi ek mærð snara,

þengil mun þess vara,
þat nam ek lítt spara;
finnrat frœknara
fœði gunnstara,
mann né mildara,
merkir blóðsvara.

Þessi er ok full runhenda. Þessi er hin minni runhenda †ok tekin af hálfhneptum hætti eða náhendum:

Þengill lætr hǫpp hrest, **93**
honum fylgir dáð mest;
vísi gefr vel flest
verbál ólest.
Húfar brutu haf ljótt,
heim *lét ek jǫfur sótt,
yngva lofa⟨r⟩ ǫll drótt,
jarl⟨s⟩ sá *ek frama gnótt.

Þessi er stýfðr eða hneptr af fyrra hætti.

[. . . . .] gramr, **94**
gull⟨i⟩ søri Kraki fram⟨r⟩,
efla frágum Haka hjaldr,
[. . . . .] aldr;
ormi veitti Sigurðr sár,
slíkt var allt fyr liðit ár,
Ragnarr þ[ótti] skatna skýrstr;
Skúli jarl er myklu dýrstr.

⟨Málaháttr:⟩

Munða ek mildingi, **95**
þá er Mœra hilmi
fluttak fjǫgur kvæði,
fimtán stórgjafar.
Hvar viti áðr orta
með œðra hætti
mærð of menglǫtuð
maðr und himins skautum?

⟨Fornyrðislag:⟩

96 Ort er of ræsi
þann er rýðr granar
•vargs ok ylgjar
ok vápn litar;
þat mun æ lifa
nema ǫld farisk,
bragni⟨n⟩ga lof,
eða bili heimar.

⟨Bálkar lag:⟩

97 Lypta ek ljósu
lofi þjóðkonungs,
upp er fyrir ýta
jarls mærð borin;
hverr muni heyra
hróðr gjǫflata
seggr svá kveðinn
seims ok hnossa?

Sú er grein milli þessa hátta at í fornyrðislagi eru í fyrsta ok þriðja vísuorði einn stuðill, en í ǫðru ok fjórða vísuorði þá stendr hǫfuðstafr í miðju orði; en í stikkalagi eru *tveir stuðlar en hǫfuðstafr í miðju orði, en í Bálkar lagi standask stuðlar ok hǫfuðstafr sem í dróttkvæðu.

Starkaðar lag:

98 Veit ek verðari
þá er vell gefa,
brǫndum beita
ok búa snekkjur,
hæra hróðrar
en heimdrega
—unga jǫfra—
en auðspǫruð.

99 Þeir ró jǫfrar
alvitrastir,
hringum hæztir,
hugrakkastir,
vellum verstir,
vígdjarfastir,

hirð hollastir,
happi næstir.

Ljóðaháttr:

Gløggva grein 100
hefi ek gert til bragar,
svá er tírœtt hundrað talit;
hróðrs ørv⟨er⟩ðr
skala maðr heitinn vera
ef sá fær alla háttu ort.

Galdralag:

Sóttak fremð, 101
sótta ek fund konungs,
sóttak ítran jarl,
þá er ek reist—
þá er ek renna gat—
kaldan straum kili—
kaldan sjá kili.

Njóti aldrs 102
ok auðsala
konungr ok jarl.
Þat er kvæðis lok.
Falli fyrr
fold í ægi
steini studd
en stillis lof.

Textual Notes

0/1–3 *Heading from U; R has no legible heading, but begins* 0/4 *with a large decorated capital. In T the heading is* Upphaf Háttatals. *The beginning of* Háttatal (*to 6/10*) *is lacking in W* 0/15 grein *TU*, málsgrein *R* 0/23 *written* háttanda *R* 0/28 greinir (2)] *so U, written* grein er *RT*

1/1 *Heading in U**: Fyrst er dróttkvæðr háttr, *in U*: Dróttkvæðr háttr. i., *in T*: It fyrsta kvæði 1/7 á *T*, of *RU* 1/23 *Unlike T, R has a stop after* kveðr, *instead of after* at 1/24 á *is a conjecture*: *RT have* o, *U has* io (*and lacks* í *but adds* ok) 1/31 *supplied from T* 1/44 *Heading in U*: Kendr háttr. ij. 1/55 *begins with a large capital in R*

2/1 *There are traces of red, but no legible heading in R* (*read* Kendir hættir *in* SnE (*1931*), Þetta er kent *in* Skj *A ii*; *probably belongs at 1/55*). *Heading* ii. kenningar *T*, Kendr háttr *U** (*cf. 1/44*) 2/3 fang] *written* fa/ng *R*, faung *T*, favng *U* 2/6 ok] *added by R**; *also in T*; en *U* 2/8 -seið] *a letter written after the* s (n*?*) *deleted R* 2/11 *written* renningar *R* 2/13 fúr *TU*

3/1 *R has a trace of red but no legible heading*; Rekit (*numbered* iii) *TU*, Rekit *U**

4/1 *Heading*: Sannkent *U**, Sannkent. iiij. *U* 4/8 bjartr *TU* 4/18 *supplied from TU* 4/21 *Traces of a red heading R* (þetta er tvíriðit*?*) 4/22 *RT add* þetta er tvíriðit kallat, *but no example follows. Possibly an explanation or exemplification has been lost from the text*

5/1 *Heading in R not legible*; *in U (numbered* v.) *and U**: Tvíriðit; *in T*: Stuðningar v. (*which seems to be correct*) 5/7 vig- *R*, veg- *R** -hrœsinn] *first vowel written* o *RU*, ø *R**, ǫ (*with the hook above the* o) *T* 5/8 vald *RU*, valdr *T* 5/10 en (2) *altered in R*

6/1 *Heading in R not legible*; Nýgervingar *U* (*numbered* vi.; *spelt* -gjǫrv-) *and U**, Nýgjǫrningar vi. *T* (*which uses this form consistently in the text too*) 6/7 at *R** (*?*) *and UT*, ór *R* 6/14–15 *corrected from WU* (*U has* sverðit, kallat); svá sem sverð (*written* seerð *R*) sé ormr kallaðr, fiskr *RT* 6/17 *corrected from T*; dróttkvæðum hætti *R*, dróttkvæðr háttr *WU* 6/22 *R has a large capital*

7/1 *Heading*: Oddhent *U**, Oddhent. vii. *U*, vii *T* 7/2 *written* -hlæs *R* 7/3 hann *R* and TWU*, hér *R* 7/5 ýgr] *TWU. R has* yggr, *which is possibly an adjective, but a long vowel is to be expected for the rhyme* (*cf. 16/14, 38/11, and st. 58 nn.*), *and the scribe does not consistently distinguish long and short consonants* lætr *UW* (*though in W the vowel is* *unclear*), hætr (*written* hetr) *T*; *R has* etr *with* h *added above the line*; *this is altered to* hvetr *by R** 7/9 *supplied from WU* (*lacking also in T*)

8/1 *Heading in U*: Ǫnnur oddhending (*?—nearly illegible*), *in T*: viii. (*no title to this verse in U**) 8/2 *the* -r *added in R**; *also present in WU* (*in T* hiara reg *is written for* hjarar egg) 8/3 *the* e *in* skekr *added by R** 8/8 *written* þra/mu *R* (*?*), *altered to* þrymu *R**; þrumu *TW*, þrimu *U* 8/19 tveim] *altered to* tveimr *R** 8/22 Varðak *T*, Varðat *R* (*the couplet is lacking in WU*) 8/23 þorði *T*, þorðu *R* 8/32 stuðill] *All four MSS place lines 38–40* (sjá) *here* 8/37 -gerð *TW*(-gjǫrð)*U*, -grið *R* 8/41 *R has a large capital*, *W opens a new paragraph*, *U has a rubricated heading* Hér segir af sextánmæltum (*abbreviated* -mlm *with superscript stroke*) 8/44 er *WTU*

9/1 *R had a red heading here*: Sextánmæltr (*?*); ii. háttr *T*, Sextánmælt *U** 9/5 skekr *R*TWU*, skefr *R* 9/11 *Trace of a heading in R*.

10/1 *Heading in U**: Áttmælt, *in U*: Áttmæltr háttr, *in T*: iii. háttr 10/3 skerr *WU*, sekr *R*, skekr *T* (*cf.* 9/5)

11/1 *Heading in TU**: Fjórðungalok (*T adds*: iiii. háttr)

12/1 *Heading in RTU**: Stælt (*T adds*: v. háttr) 12/1, 10 ok *altered from* en *R* 12/4 -konungs *R*WT*, -konungi *R*, konungr *U* 12/5 Vald *RWU; altered to* Va/ld *?R**, vauld *T* 12/9 *supplied from W, RT omit*; Þetta er it fyrsta *U* 12/11 *R and T repeat ll. 1–2 of the stanza instead of ll. 1 and 4, as in WU* (*which here has* konungs)

13/1 *Heading in R*U**: Hjástælt, *in T*: vi. háttr 13/5 gat] *appears to have been written* gatk *or* gack *in R* (*?*) *and altered to* kann *by R**(*?*) 13/9 *supplied from WU*, *lacking in RT* 13/11 hinu *W*, inu *U*; hina *R* (*the phrase is lacking in T*) er eptir fara *WU*, eru eptir vísuorð *R*, en þær eru eptir vísuorð *T*

14/1 *Heading in R*U**: Langlokum 14/9 í *WU*, *lacking RT*

15/1 *Heading in R**; Tiltekit, *in U**: Afleiðingum 15/3 þann] *conjecture*; þanns *R*, þats *WT*, þess *U* 15/5 -hættir *R* and TWU*, -heitir *R*

16/1 *Heading in R*U**: Drǫgur 16/4 aldir *R*TWU*, aldri *R* 16/7 þryngr *R*TW*, þungr *R*, þraungr *U* 16/9 *corrected from W*; síðars er *R*, síðarst er í *T*, síðast var í *U* 16/10 síðari *U*, síðara *RT* (*clause omitted in W*) 16/11 *begins with a large capital, and there may have been a heading R*; *heading in T*: Refhvǫrf. x. háttr, *in U* (*rubricated*): Hér segir um refhvǫrf 16/13 vera *WU*, *lacking RT* 16/15 En *TWU*, er *R*

17/1 *Heading in U**: Refhvǫrf 17/5 válkat] *written* válkar *R*(*?*) 17/6 rǫðuls *U*, rǫðul *RTW* (*cf. 17/23 and 38*) 17/11 Annat] *unclear; perhaps* An[nat] *R* 17/13 eitt fall *TWU*, ætt full- *R* 17/14 *written* Þriðu *R* 17/16 *supplied from WU*, *lacking in RT*

17/19 skilr] *R adds* fylkir, *deleted* 17/32 rǫðuls] -s *added* (*or altered from* -l) *R* 17/34 *Heading in U*: A[nnat] refhvarf, *in T*: xi. háttr

18/1 *Heading in U**: Ǫnnur refhvǫrf 18/5 lætr of] *conjecture*; látit *RTWU* (*cf. 18/18*) 18/17 skelfr] *lacking in all MSS* í *WU*, *lacking RT* 18/19 er *TWU*

19/1 *Heading in U**: In iii. refhvǫrf, *in T*: xii. háttr (*and similarly in the succeeding stanzas*; *from st. 26 omitting the word* háttr, *from st. 30 with arabic numerals* 19/4 unna *TWU* 19/5 *written* v *with superscript* i *R* 19/11 En í] *supplied from WU* (*though these lack* inum); *lacking in RT*

20/7 þyrr *could be read* þvæ *RT*; þurr *U*, *unclear W*

21/1 *Heading in U**: Ǫnnur in minni 21/2 skut *TWU*, skot *R* 21/4 Róða *WT*, rjóða *R*, *altered to* róða *R**, bjóða *U*

22/1 *Heading in U**: In þriðja 22/4 -drasill *R*, *altered to* -drasil *R**, drasil *W*, *written* drûsil *T*, drasill *U* 22/7 kløkkvan *R*TWU*, kykkvan *R*

23/1 *Heading in U**: Refhvarfa bróð⟨i⟩r 23/6 it *U*, at (réttu) *T*, hin *W*, in *R* (*written* en) 23/11 í *T*, *lacking RU*; *W reads* ok eiga eina tíð 23/14 *R has a large capital here*; *T opens a new paragraph with the heading*: Dunhent. xvii háttr, *U has a new chapter with the rubricated heading*: Hér segir hversu skipta skal hættinum

24/1 *Heading in R**: Dunhenda, *in U**: Dunhent (*cf. T at 23/14*)

25/1 *Heading in U**: Tilsagt

26/1 *Heading in U**: Orðskviðuháttr (*written* Orþr-)

27/1 *Heading in U**: Álagsháttr skar *R*TU*, skal *R*, braut *W* 27/3 fundar *TWU* (funda *added in a later hand R*) 27/16 Þetta *has a large capital in R*

28/1 *Heading in U**: Tvískelft 28/4 skipreiðu *TW*, -ræði *U*, -reiðum *R* 28/7 stál- *R*TWU*, stol- *R* 28/9 er *TWU* 28/14 *Large capital in R*

29/1 *Heading in U**: Detthent 29/2 -vistir *written* vist er *R* 29/6 flen- *R*, flein- *R*TWU* 29/10 Þetta *has a large capital in R*

30/1 *Heading in U**: Draugsháttr

31/1 *Heading in U**: Bragarháttr støkkvi *R**, støkki *R* 31/4 feng *R*TWU*, fengs *R* 31/7 *written* saltaunnu *R*

32/1 *Heading in U**: Liðhendum 32/10 *supplied from T*; í einum staf *W*, báðar í einn hljóðstaf *U* 32/11 sé *TWU*, sem *R*

33/1 *Heading in U**: Veggjat 33/10 Nú *with large capital R*; *could be read* Ne

34/1 *Heading in U**: Flagðalag 34/2 fagrskjǫlduðustum] *final* -m (*nasal stroke*) *unclear R* 34/6 harðsveipaðastan] *first* s *unclear*; *perhaps* i *R*(?)

35/8 -láð *R**TU*, -ráð *R*, -lið *W* 35/10 at *TWU*

36/1 *Heading in U**: Þríhent 36/5 herjum *R***TU*, hverjum *RW*

38 *This stanza is written at the end of* Háttatal *in R* (*where it is followed by* Jómsvíkingadrápa), *after st. 54 and its commentary in W, and not at all in T* (*which lacks the end of the poem*). *It is found in this place only in U. In W it is this stanza that is said to be in a metre invented by Veili* (*35/10–13*); *there is no commentary on the stanza in R or U, 38/9–10 are only in W* (*after the lines about Veili*). *The words* í fyrsta ok þriðja vísuorði *in W are deleted, the second syllable of* skothend *is abbreviated, and* in *is written* enn 38/1 *is thus in WU; in R it is written* Snarar farar arar snarar fylkir byrjar, *with* arar snarar *deleted by R** *and a mark indicating that the order of the first two words is to be reversed added* 38/3–4 *inserted at the end of the stanza R* (*probably not R**) 38/5 sýjur] *written* sygior *R with the* g *deleted by R** 38/7 Haka *written* Hvaka *R and the* v *deleted by R**

39/5 foldar *R***TWU*; *apparently written* folkar *or* folrar *R*

40/3 gerir *unclear in R* 40/6 blikur *TU*, bliknir *R*, blikur *?R**, blikurs *W* 40/9 Þessum hætti *TU*, Þessi heiti (*or* hetti = hætti) *R*, Í þe[ss]um hætti *W*

41/3 þrøngvi *R***TWU*, þrøngi *R* 41/7 *written* hættiz (*i.e.* hættisk*?*) *R* 41/11 inu *T*, einu *R*, *lacking U*; fyrra *U*, *lacking RT* (*clause lacking in W*)

42/3 hvar *TWU*, hvat *R*(*?*) 42/11 alhendu *U*, aðalhendu *RT*, *sentence lacking in W*

43/1 søkkum *TWU*; *written* søkkv *R*, søkkvi *R** 43/7 -vinjat *RTU*, -vinjaðr *R**, -vinjar *W*

44/1–18 *come before 43/1 in W, which is the more logical order* (*W omits* 42/11–13 *and* 43/12 *and* 44/19) 44/6 hagbáls] -s *deleted R**(*?*)

45/5 blik *R***TWU*, brim *R*(*? perhaps* briu *or* brici) 45/9 *supplied from TWU*

46/2 verbál *TWU*, verbáls *R* 46/4 afar *TWU*, aurar *R* 46/6 auðs] -s *deleted R**(*?*) 46/10 fyrsta *W*, ǫðru *RTU* 46/12 *R has a large capital*

47/4 þinga *RTU*, stinga *R**, -ninga *W* 47/6 drauga *WU*, draugar *T*, draugum *R** 47/9 samhending *TWU*, samhendingin *R* 47/12 *Large capital R*

48/1 Auðkendar *TU*, Auðkendur *R*, Auðkendum *W* 48/3 þar *R***WU*, þat *RT* 48/7 venr *TWU*, verr *R?* 48/11 *R has a large capital*

49/9 stýft] *R adds* tekin

50/2 móð *R***T*(*? or* með)*WU*, *written* mð *R*

51/9 eru *U*, ok *RT*, *W omits* 51/12 ǫll *WU*, *RT omit*

52/4 of *T*, af *RW*, um *U* 52/4–6 *supplied from TWU* (*U has* fyrir *for* fyr; *the word is abbreviated in W*)

53/3 kann] *perhaps* hann *R* 53/11 *Heading in U*: Hættir fornskálda

54/1 *Heading in T*: Ragnars háttr (*and numbered* 46; *R may also have had the numeral* xlvi *here, but it is no longer legible*) 54/7 vinr] *the* -r *has possibly been erased in R* (*by R*?*) 54/9 er *TWU*

55/1 *Numbered* 47 *in T; the number in R is no longer legible* 55/4 hraðsveldan] s *deleted in R, altered to* hraðfeldan *R** 55/5 af *TWU*, of *R* 55/7 jaðarr *WU*, jarðar *RT* 55/10 Nú] *large capital in R*

56/1–10 *after 57/10 in W* (*which omits* Nú . . . háttr *in each case*) *Stanza-number no longer legible in R* (*numbered* 48 *in T*) 56/7 herr *altered to* hann *R** 56/10 *The text of U ends here* Nú] *R has a large capital*

58/5 kná *R* TW*; *R appears to have* búa *or* bna 58/11 inu *T*, *written* ena *R*, *W omits* 58/12 *The words supplied are editorial conjecture*; *fo.* 50^r *in R ends with* lið, *which may be an abbreviation for* liðhending; *the top line of fo.* 50^v (*which is nearly illegible*) *seems to begin with* ina fyrstu. *T has* skothending ok liðhending við ina fyrstu; *W omits this part of the sentence* 58/14 vísuorði] *RT add* er 58/14, 16 Víða er þat í . . . því] *supplied from TW*

59/7 Glǫndlar *R* 59/8 gagn *TW*, gang *R*

60/3 hræs] *so W*, hress *T*, hre[. .] *R* (*with an alteration that is not legible*; SnE (*1931*) *reads* hvatt) 60/7 frár *R**; *written* fraʋr *R*, frǫr *T* (*with the hook above the* o), fǫr *W*

61/1 fellr *R**, felr *RTW* 61/3 hlyn *R*W*, hlynr *RT* 61/8 skotnar *has been altered* (*by R*?*) 61/10 *The text of T ends here*

62/13 *supplied from W* 62/16 hrynhenda] *altered to* hrynjandi *R* (*apparently in the same hand*)

63/1 kendi *R*W* 63/9–10 *added at foot of page with indicative sign R* 63/10 þrjár] *emendation*; fjórar *R*, *W omits this part of the sentence*

64/3–8 *supplied from W*

65/7 *corrected from W*; ǫflum sótti oddi *R* 65/9–10 *supplied from W*

66/1 *Rubric in R*: lviii 66/9 *corrected from W*; fyrstu *R*

67/11 í *supplied by editors*; með *W* 67/14 *R has a large capital here, W opens a new paragraph*

68/8 hers *R*W*, hans *R*

69/9 annat *W*, annan *R* 69/12 *supplied by editors*; *the sentence is omitted in W* (*for the name* hagmælt *cf.* 71/9)

70/1–7, 9 *supplied from W* (*though there* skothendingar *is abbreviated* skoth. *and the verb at* 70/9 *is singular*)

71/1 *R has an illegible rubric* sneiðir] *R**; *R seems to have* snæðir 71/7 leyfðr *seems to have been altered in R* 71/11 endask *W*, endar *R*

72/1 *Rubric in R*: lxv kná er *R*, *altered to* knáir *R** (*i.e. the two words are joined by a line*; *this may be intended as a 3rd pers. sg. form*)

73/1 *Rubric in R*: lxvi 73/2 døkkva *W*, støkkva *R*, *altered to* døkkva *R** 73/10 aðalhendingar] *abbreviated* aðalhend. *R*

74/1 *Rubric in R*: lxvii 74/7 bǫðharðr bǫrðum *W*, bǫðhjarðr hǫrðum (*altered to* bǫrðum *R**) *R*

75/1 *Rubric in R*: lxviii 75/2 snekkjur *altered to* snekkjum *R** 75/5 hergramr *W*, her fram *R* (*possibly an error for* herframr) 75/7 raungóð *W*, rauð góð *R* 75/9 *supplied from W*

76/1 *Rubric in R*: lxviiii 76/11 skothending] *abbreviated* skothend. *R*

77/1 *Rubric not legible R* 77/1–5 *supplied from W* 77/2 hlekk] *perhaps* hlokk *R* 77/11 aðalhendingar] *abbreviated* aðalhend. *R*

78/10 *supplied from W*

79/1 *Rubric in R*: lxxii 79/3 *corrected from W*; *R** *seems to have* um bláskoðu 79/6 val-] fal- *R*W* 79/7 élsnœrðum] *the second vowel written* o *R* 79/11 *Rubric*: lxxiii *R*

80/6 *R** *adds* til *before* hróðrs; *W reads* til hróðrar ǫrum] *written* a/rum R, *altered to* geyrum (*i.e.* gørum) *R**, gjǫrum *W*

81/1 *Rubric in R*: lxxiiii (?) 81/2 of] *emendation*; ok *R*, um *R*W* 81/9 *corrected from W*

82/1 *Rubric in R*: lxxv 82/4 orðróm *R*W*, ǫðrum *R* 82/9 *supplied from W*

83/4 -fjalla *R*W*, -fjallar *R* 83/5 galla] *altered from* gjalla? *R* (*or R*?*) 83/6 gumna] *altered from* gunna *R?* 83/7 fremr] *written* freimr *R*

84/1 *Rubric illegible R* 84/9 *supplied by editors* (*according to Finnur Jónsson*, SnE (*1931*), *249*, fyrri *is written in the margin*); ok með minni runhendu (*abbreviated* runh.) *W*

85/2–3 *and* 8 *supplied from W*

86/3 er *R*W* 86/5 drekkr *R*W*, dregr *R*. *The text of W ends with this line* 86/11 *written* hrunhenda *R here and at 88/9*

88/1 *Rubric illegible R* 88/2 hers *R**, her *R*

89/1 *Rubric illegible R* 89/10 runhenda] *abbreviated* runh. *R*

90/1 *Rubric illegible R* 90/3 seima *R** 90/9 runhenda] *abbreviated* runh. *R*

91/1 *Rubric illegible R* 91/9 runhendu] *abbreviated* runh. *R*

92/1 *Rubric illegible R* 92/9–10 ok tekin . . . náhendum *written after* runhenda (1) *in l. 9 in R*, *but seems to belong to the description of st. 93* 92/10 -hneptum *written* hnefstum *R*

93/6 leit *R* 93/7 lofar *R**, lofa *R* 93/8 jarls *R** (?), jarl *R* ek] er *R*

94/2 framr *R**, fram *R* 94/9 *Heading added by R**; *similarly 95/9, 96/9*

97/10 vísuorði] *R adds* er 97/11 tveir] þrír *R* 97/13 *Heading in margins R (where it apparently relates to st. 99) and R**
99/9 *Heading added in margin R*
100/4 -verðr *R** 100/7 *Heading in margin R*
102/8 *R has st. 38 here without any commentary, see textual note above; then follow* Jómsvíkingadrápa (*beginning at fo.* 53^r29) *and* Málsháttakvæði

Explanatory Notes

0/1. Háttatal: 'enumeration of metres'; though since some of Snorri's varieties are stylistic or grammatical rather than metrical, perhaps 'enumeration of verse-forms'. *Hættir* at 0/4 seems to be used in an even broader sense, in view of the answer given at 0/7. Cf. also 95/6. At 1/25 and in the term *háttlausa* (see Glossary) it is internal rhyme (or the lack of it) that is referred to; cf. 58/14–16 and n., and Glossary under *tøgdrapulag*.

0/7. These three categories are clearly based on the traditional *pars praeceptiva*, *pars permissiva*, and *pars prohibitiva* of Latin grammarians. The first normally referred to Donatus' books 1 and 2 (orthography and grammar or morphology), the second and third to the figures of speech (those that were permitted and those that were considered defects of style) in book 3. On *leyfi* cf. *GT Prologue* 155.

0/11. Rétt ok breytt: 'normal and varied'. Compare the opposition *naturalis*~*artificialis*, applied to rhetorical ordering of discourse in a direct or a roundabout way in Fortunatianus, *Ars rhetorica*, 3. 1 (Halm 1863, 120). Alcuin (1941), 100 has the exchange: 'An semper perspicue exordiri debet orator?—Aliquando perspicue, aliquando per circuitionem.' The terms *naturalis* and *artificiosus* are applied in the *Rhetorica ad Herennium*, 3.16.28 to *memoria*.

0/16: i.e. *réttrar setningar*. Similarly at 0/23.

0/19. tala: another term derived from the terminology of grammar. The three categories into which Snorri divides it are illogically on three different levels of discourse and are not complementary to each other.

0/29. harðar eða linar: presumably referring to quality of sound as opposed to length, though it is possible that the terms refer to accented and unaccented syllables. Accentuation, however, is less relevant to rhyme, which seems to be the topic under discussion here. Cf. *GT Prologue* 154.

Stanzas 1–30 are in praise of King Hákon, 31–98 are predominantly in praise of Earl Skúli; the last few stanzas praise the two rulers together. Finnur Jónsson (1920–4), ii. 79 and de Vries (1964–7), ii. 79 both strangely claim that stt. 68–101 are predominantly about Hákon, but there is no unequivocal mention of the king after st. 67 until st. 97.

1/9–16. Cf. *TGT* 96–7, where alliteration is also discussed. *Kveðandi* seems here and in some other places to be used to mean 'poetical effect' in the most general sense, but elsewhere it seems to refer particularly to either alliteration (e.g. 1/13) or rhyme (e.g. 1/26).

1/40–1. orð here as frequently elsewhere = *vísuorð*. The objects of *kǫllum* in these two lines may be understood (i.e. *frumhending*), but possibly *þá* is the fem. acc. pron. rather than the adverb.

1/44. The writer distinguishes three kinds of variation of form (*háttr*): variation in sound, *breytt með hljóðum* 1/47, 27/14 (i.e. in arrangement of rhyme or alliteration or in length of line), stt. 28 onwards (though st. 24 seems to belong with these, cf. 27/13 n.); rhetorical variations, *með máli skipta* but *halda sama hætti* 1/48, 52 (systematic use of kennings etc.), stt. 2–6; and syntactical variations, *breyta háttum með máli einu* 8/45 (end-stopped lines, various kinds of parenthesis, linking of stanzas etc., and use of antithesis), stt. 9–23 (and 25–7). Stt. 7–8 involve variations in length of line due to various kinds of elision and resolution, which he also does not seem to think of as 'real' variations (cf. 67/10–12).

2/5–8: quoted in the additions to *Skáldskaparmál* in W to illustrate kennings for the head (*SnE* 1924, 111).

2/10–11. As elsewhere one of the subdivisions of the category (*kenning*) has the same designation as the class to which all three subdivisions belong. Presumably the first category should be understood as 'simple kenning'; cf. 4/21 (*sannkenningar*). These passages, which seem to have been written before the accounts of the kenning in *Skáldskaparmál*, chs. 1 and 67, are the clearest indication that *Háttatal* was the first part of the *Edda* to be composed. Snorri's analysis of the kenning was refined after the writing of the commentary to *Háttatal*.

Stanza 3 seems to be exemplifying the systematic use of *tvíkent* (two examples in ll. 1–2, one in l. 5, one in ll. 7–8); *rekit* (when kennings have more than two determinants) is not exemplified in a separate stanza, though examples appear in 2/1–2 and 3/4. Cf. textual note on 3/1.

4/8. E. A. Kock (*NN* 2174) points out that the pattern of this stanza requires an adjective with each noun and an adverb with each verb, and that R in fact has the neuter form *bjart*, which could be taken as adverbial (with *unir*; though it does not give easy sense as such). On the other hand the writer's paraphrase of this line at 4/17 uses the masculine adjective *glaðr* as equivalent of *bjartr* (which is the reading of T and U at 4/8), and an adjective in agreement with the subject in such a context is virtually adverbial in force.

4/18–20. An acknowledgement that Snorri has systematized a variant (using it in every line when one would expect it to be only an occasional embellishment), thus making a new 'form' from it. Cf. 6/18–21.

4/21–5/11. This passage was used by the reviser of *Skáldskaparmál* in W, see *SnE* (1924), 105 (cf. also 6/20). The confusion noted in 4/22, textual n. reappears there; perhaps the mistake arises from separating *stuðning* and *tvíriðit*. Stanza 5 illustrates the systematic use of *stuðning* (though apparently only in ll. 1–4: the compounds in the second half of the stanza are of a different kind), and there is no stanza illustrating *tvíriðit*.

5/3–6: quoted in *TGT* 107 in illustration of *epitheton*.

6/9. The meaning of *nýgjǫrvingar* in *Skáldskaparmál* is slightly different. It is there used to refer to extensions of meaning by the use of near-synonyms even if the result does not involve metaphor; that is, it means the making of new kennings on the pattern of older ones (cf. especially *Skáldskaparmál*, ch. 33). According to Hallvard Lie in *KLNM* ix. 560 ('Kviðuháttr'), *nýgjǫrvingar* are particularly frequent in *kviðuháttr* poems (see note to st. 102).

6/15–16. Egill Skallagrímsson sometimes used *nykrat*: *lausavísa* 23 (*ÍF* ii. 172) 'Þél høggr stórt fyrir stáli'; *Hǫfuðlausn* 8, 17 (*ÍF* ii. 188, 191); cf. also *TGT* 80, *FoGT* 131. Snorri's comments on mixed metaphors have been compared with Quintilian, *Institutio Oratoria* 8. 6. 50 (1920–2, iii. 328–31), but the similarity, though striking, is not sufficient to indicate derivation. On *nykrat* in Norse poetry in general see Macrae-Gibson (1989), 170–1 and references in nn. 45–6.

6/17. dróttkveðinn is the reading of T, and is the form likely to have given rise to the erroneous reading in R; but this is the only occurrence of this pp. in the text, and W and U have the more usual *dróttkvæðr*.

6/17–18. hinn sami háttr: i.e. stt. 2–6 are not really in a different verse-form from stanza 1. Cf. 1/52 and 1/44 n., 4/18–20 n.

6/24. í ǫðru ok inu fjórða vísuorði: i.e. of each half-stanza, as throughout the commentary to *Háttatal*. Similarly 8/9: *í fyrsta ok þriðja vísuorði*. Cf. Glossary under *vísuorð*.

Stanza 7: all four of the five-syllable lines (and one six-syllable one, l. 3) contain contracted words which at an earlier stage of the language would have had an additional syllable: *-hlés*, *kná*, *hræs*, *-ár*, *-fá*, *-grá*. Though the author does not say this, it is likely that such contractions gave rise to short lines and made the lack of a syllable acceptable, though it is also likely that the extra syllables continued to be used in performance of poetry long after they had disappeared from ordinary prose speech. Cf. Kuhn (1983), 69–70. Eilífr Guðrúnarson, *Þórsdrápa* 5/8 (*Skáldskaparmál*, ch. 18) may have been Snorri's model for st. 7; as elsewhere he has made a new verse-form out of an occasional phenomenon (cf. Bjarni Einarsson 1971, 22–3; 1987, 151–3). Note also Kormakr's *lausavísa* 18/8, *ÍF* viii. 222. Similar phenomena occur in eddic verse, e.g. *PE Vǫluspá* 42/4, *Guðrúnarkviða* I 13/4, 19/6, 26/8.

Stanza 8. One of the extra syllables in each line except l. 3 is accounted for as providing resolution of a stress with a preceding short syllable, in each case in the first word of the line. Other additional syllables in the odd lines would disappear if pronouns were suffixed (*spyrk* in l. 1) and conjunctions were contracted (*þvíro*, *þars* in l. 3, *þótt* in l. 5) and if the monosyllabic forms *fyr* and *of* were used for the prepositions *fyrir* and *yfir* (ll. 1 and 7). Lines 5 and 7 have further resolutions (*muna* and *þorir*). Line 3 still has an extra syllable, and this can only be got rid of by making the first verb singular, as in W, as well as eliding its vowel (reading *þvís*). It may be therefore that this stanza actually contains no 'additional' syllables in reality.

Snorri does not clearly show that he understands the principle of resolution. Contraction and elision he describes below as *bragarmál*. In manuscripts many scribes have expanded such contractions (just as they may have contracted such words as account for the short lines in st. 7), and it was maybe this that led Snorri to think that the extra syllables could have been pronounced without spoiling the metre, though it is likely that most readers would not have pronounced them, but would automatically have elided them whether they were written or not. Skaldic stanzas rarely contain extra syllables that cannot be accounted for in one of the above ways, but whether Snorri believed that extra unstressed syllables were acceptable in *dróttkvætt* as they were in eddic metres even if they were neither for resolution after short stressed syllables nor capable of elision is difficult to say. The remark at 8/12–14 suggests that he thought of the extra syllables as being a prelude or anacrusis. Since the initial syllable in each line (except l. 3) is short, the hending would not normally fall upon it anyway, cf. 38/10 n. and Kuhn (1983), 80–2, where examples from 10th-c. verse are given; hendings with short syllables were avoided by later poets, see Kuhn (1977). On the positions where resolution was acceptable in *dróttkvætt* see Kuhn (1977); (1983), 55–6, 68–9; for examples of resolution and elision see Sievers (1878).

8/8. The reading of R is uncertain (see textual n.), though Finnur Jónsson read *þrǽmu* in *Skj* A ii. 54 (*þrumu* in *SnE* 1931) and adopted *þrǫmu* 'edges' in *Skj* B ii. 63; but the use of a kenning might be expected here (*þrumu randa* = battle). *Hreinn* is an odd adjective to use with it; perhaps it could be translated 'pure'. W and T have *hreinna* (gen. pl. with *randa*).

8/10. hversu flestar: see Heusler (1950), § 394.

8/13. It is uncertain whether *þat* refers to the preceding clause or the following one in l. 14 (in the former case the second comma in l. 13 should be omitted and *at* interpreted as 'so that').

8/16. með álǫgum: cf. *álagsháttr*, st. 27.

detthent: st. 29.

dunhent: st. 24 (cf. textual n.).

skjálfhent: cf. 28/11 and st. 35.

8/17–18. Cf. *rétthent*, st. 42.

8/18–20: i.e. *varðak* for *varða ek*, *þars* for *þar er*. These are usually taken to be archaisms rather than poetical elision. Cf. Noreen (1923), §§ 158, 465, 473 Anm. 2; Einar Ólafur Sveinsson (1958), 97 and 101.

8/22–3. The stanza from which these two lines (only present in R and T) come is in *Eyrbyggja saga*, *ÍF* iv. 38 (*Skj A* i. 111, B i. 105). The sentence as quoted is incomplete: the indirect object of *varðak* is *kvinna frýju* ('from women's reproach'). *Myrðir morðfárs* refers to the poet.

8/24. sú grein: i.e. *leyfi*.

8/24–5. skipta tíðum: cf. 8/41–4 and *TGT* 77.

8/26. samhendingar, liðhendingar: see stt. 46–8 and 41, 53, 58.

8/29. vísuorð eða skemra: apparently 'whether it be [repetition] of a whole line or less'. Repetition of words within a stanza is normally not permitted (cf. 8/32–40), but may be an acceptable *leyfi* if it is effective (*at nýta*).

8/29–31. Cf. *LP* s.v. *gimsløngvir* (the stanza quoted is in *Hkr* i. 187).

8/31–2. drag eða stuðill: both unexplained; it is not even certain whether these are two terms for the same thing. *Stuðill* here is clearly different from *stuðill* as applied to alliterating staves (1/14 etc.). *Drag* may mean an extra line added at the end of a stanza (when this occurs it usually repeats or echoes l. 8; cf. *galdralag*, see st. 101 n.), or could be the same as *drǫgur* (st. 16). There are various examples of the former quoted from dream-verses in Möbius (1879–81), ii. 129 (including *Skj* B i. 400, st. 10, which is in *Hkr* iii. 177; see also *Skj* B ii. 609 under *varselsvers*). Cf. also *Háttalykill*, st. 22 (*núfuháttr*), which has a refrain-like addition to each quatrain. Snorri seems to confine the term *galdralag* to *ljóðaháttr* with an extra line at the end of the stanza (st. 101), while the dream-verses referred to above often add the extra line to *dróttkvætt* stanzas.

8/33. í vísuhelmingi: *í vísuorði* would make better sense (*er* comes twice in 8/34).

8/34–7. Not recorded elsewhere; taken to be part of Refr's poem about a certain Þorsteinn (perhaps the son of Snorri goði). For Hofgarða-Refr's extant verse see *Skj* A i. 318–21 (it is only preserved in manuscripts of the prose *Edda*, *Heimskringla*, and *Óláfs saga helga*).

8/38–40: i.e. to repeat a reference to an individual previously named or referred to, even if only by the use of a pronoun, is counted as *leyfi*. This clause presumably belongs with the seventh *leyfi* above (8/26–8). Cf. textual note on 8/32.

8/45. með máli einu: i.e. syntactical variation, cf. 1/44 n. Here Snorri introduces the first of what he considers to be real variations of form, and from now on most variations are given a name (8/46–7).

8/46–9. Cf. 0/19–22.

11/9–10. Stt. 2–8 were not considered to be fully *breytt* (cf. 1/52), since the variations in them were only rhetorical, or else involved *leyfi* (stt. 7–8). Stanza 12 is therefore the fourth variant verse-form (st. 10 was the second, cf. 9/11) and the fifth stanza-type including the standard type of st. 1. Stanza 13 is the fifth variant (12/13), stt. 14–17 are respectively the seventh to tenth verse-forms (taking st. 1 as the first).

Stanzas 12 and **14** are quoted in *FoGT* 136–7 in illustration of *antitheton*.

13/4, 8. The two 'traditional statements' (*forn minni* 13/12) are clearly connected with an account of creation, but have not been preserved elsewhere. Cf. Ps. 104 (103): 6 and *Vǫluspá* 4, 59 (see Fidjestøl 1982, 249).

14/1. Langlokum (see textual note). The name in the dat. is elliptical, meaning '(composed) with *langlokur*', rather as some place-names were used in the dative (with a preposition), e.g. *á Hallfreðarstǫðum*, *ÍF* xi. 97. The form *langlokum* is also used in *Háttalykill*, st. 30 and *FoGT* 136. Cf. Möbius (1879–81), ii. 65 (such datives are used on other occasions in U, see textual notes on 15/1, 32/1; *SnE* (1848–87), ii. 377/23.

15/1–2. There seems to be a redundant element in the kenning: either *gjaldseiðr* or *grundar seiðr* would be sufficient to mean 'serpent'. Cf. *hranna hádýr* 8/34–5 and 60/3 n. If the kenning is to be analysed as equivalent to *seiðr gjaldgrundar* it is similar in structure to *vandbaugskaði* 86/8 and *vandbaugs sendir* 28/1 (see also *rǫðull* in Glossary and 17/6 n.). Such kennings employ a kind of tmesis, on which see 36/7–8 n. Cf. also Meissner (1921), 164.

15/9–12: i.e. *Þeim er* (15/1) is dependent on *konungdómi* (14/8). It is odd that the word *afleiðing* is applied to the latter word rather than to the first word of st. 15; *sá vísuhelmingr* means the first half of st. 15, which is not correct *tiltekit* (15/1 textual n.) unless it is syntactically linked to the preceding stanza (cf. st. 39). See Möbius (1879–81), ii. 56, where the illogicality is taken as evidence of alteration or corruption in the archetype.

16/9–10. Cf. *TGT* 94, where 15/7–8 and 16/1 are quoted as an example of *anadiplosis*. The device is used extensively in Old French and Middle English. Cf. Gordon (1953), 89, and see also JH–AH 129. Similar linking of stanzas is found in medieval Irish verse, see Heusler (1925), i. 313. Cf. st. 39 and 39/9–10 n.

16/11–15. Again Snorri seems to be making consistent verse-forms by regular use of what would normally be sporadic embellishments (cf. 4/18–20). He himself uses *refhvǫrf* as an occasional device, e.g. 44/8 and 54/3 (if there was a verb *hlemma* 'stop'; see Fritzner 1883–96, ii. 7).

16/12–13. hafi þó einnar tíðar fall: *tíð* here can scarcely mean 'tense' as it clearly does at 8/25 and 8/41 (and *TGT* 76), since many of the examples of *refhvǫrf* below concern nouns and adjectives rather than verbs. The phrase may mean the opposite of *lúkask bæði eigi í eina tíð* (23/10–11), whatever that is, and possibly the same as *í eitt fall mælt* (17/13; in neither place can *fall* have its usual meaning of 'grammatical case'). *Tíð* could perhaps mean 'rhythm' or 'cadence' (though in the first two and some other examples of *refhvǫrf* the two contrasted words are of one and two syllables respectively). The concept 'metrical foot' does not really apply to *dróttkvætt*: even if one does apply it, *verr* and *sækir* in 17/1 would not by most people be considered to be in the same foot; the phrase may, however, simply mean that the two contrasted words must be adjacent (cf. 23/10), or possibly that they must be on the same side of the caesura (in the same hemistich), cf. Kuhn (1983), 89–90 (some of the pairs of antithetical words in st. 23 are divided by the caesura, cf. 23/11). Perhaps the most likely thing is that *tíð* refers to syllabic quantity, as perhaps *fall ęðr tima* does at *GT Prologue* 154, and certainly *tíð* does in *TGT* 52–3, though for instance *hafi jarðar* (17/2) and *frǫmum meiðum* (17/8) do not fulfil this condition either. In fact the only way in which all Snorri's examples can fulfill his condition is if the phrase means that both words must be stressed. Cf. *endask báðar í einn staf* 71/11 and *lúkask á einum hljóðstaf báðar* (32/10), *lúkask allar einnig* (36/10), *lúkask í einn staf báðar* (73/10, 78/10). It is possible that the statement is meant to apply, not to the two words linked by *refhvǫrf*, but to the two words involved when there is *ofljóst*. *Ofljóst* can involve two words with vowels of different quantity, though this does not seem to happen with any of the examples in *Háttatal*, and Snorri may mean that it should be avoided; cf. *Skáldskaparmál*, ch. 74, *TGT* 172, Bjarni Einarsson (1987), 160. Similarly, hendings can link syllables with vowels of different quantity, though this was generally avoided, at least as far as *aðalhending* was concerned, cf. 38/10 n., 79/10 n. and st. 8 n. (see Kuhn 1977, especially p. 528).

16/16. frumsmíð: i.e. either this kind of verse has never been attempted before, or this is the first time that Snorri has attempted this kind of verse. Cf. stt. 70 and 95.

17/3. *reka* elsewhere in the text means 'drive, hammer' literally or metaphorically. With the object *kǫld heit* it could mean 'fulfils (cold promises or threats)', sc. to punish evil-doers (LP s.v. *heit* interprets the hostility as towards gold, i.e. indicating the ruler's generosity, but this conflicts with Snorri's own interpretation below, 17/30), but it is perhaps more likely that the threats come from the enemy and that the phrase means 'drives away cold threats' (in *Skj* B ii. 65 the verb is translated 'straffer').

17/6. If the reading *rǫðul sævar* is correct (in spite of 17/32), *rǫðul-sævar meiðum* would be equivalent to *rǫðul-meiðum sævar*; cf. 15/1–2 n.

17/13. í eitt fall mælt: perhaps 'said in the same breath, in the same phrase (clause?) or expression'; cf. 16/12–13 n. But it is difficult to see what the rest of the sentence means. Possibly Snorri has in mind that on their own *hafi jarðar* (respectively dative and genitive) could imply 'from sea to land', since the two cases fundamentally signify origin and destination (see Nygaard 1906, §§ 113–16 and 141).

17/14, 16, 24. *ljóss* may have its normal meaning of 'clear, obvious' in this passage, but it may be connected with the use of *ofljóss* at 17/26, 18/13, and 20/9, which is almost certainly being used in its ironical sense 'containing word-play, punning', as in *Skáldskaparmál*, ch. 74. *Refhvǫrf* as exemplified in stt. 17–23 nearly always requires the contrasting words to be taken in one sense to provide antithesis, but in another sense as part of the meaning of the verse.

Stanza 20. The first two words in each line give the following antitheses, usually by taking them in different senses from those required for interpreting the meaning of the verse (*flest ofljós* 20/9): freezes ~ warms; raised ~ falls; travel ~ stay; away ~ towards; twists ~ straight; wraps ~ unwraps; steps ~ rushes; warns ~ exhorts.

Stanza 21. Antitheses at the beginning of the even lines: lay ~ rose; ran ~ stood; confined ~ spacious; stiff ~ pliable.

Stanza 22. Antitheses in the first two words of ll. 4 and 8: ugly ~ fair; cold ~ embers.

Stanza 23. Antitheses in the second halves of the even lines: winter ~ summer; rest ~ movement; leaning ~ straight; empty ~ full.

23/10–11. lúkask bæði eigi í eina tíð: cf. 16/12–13 n. It is unclear whether this means the same as *standa eigi saman ok er ein samstafa millum þeira*, or whether it refers to some other difference from ordinary *refhvǫrf*. The pairs of antithetical words in this stanza all include one monosyllable and one disyllable, as well as being not adjacent and perhaps not being considered to be in the same phrase (there is a similar problem with the meaning of *lúkask á einum hljóðstaf báðar* and *lúkask allar einnig*, 32/10 and 36/10, see notes to these lines). They are all stressed words, though in some cases if the lines are taken to be Sievers type D they may be only half-stressed (this is, however, unlikely, since then there would have to be greater stress on the preceding verbs). All the main syllables are metrically long (though their length may be different according to the rules in *TGT* 52–3), but then so were most of those in the contrasted words in ordinary *refhvǫrf*.

23/11–15. But variations involving hendings do not in fact appear until st. 28, though the repetition of words in st. 24 does result in similar hendings in each pair of lines (cf. stt. 41, 47, 48, 53). Variations in length of lines do not appear until st. 33. Cf. 27/12–15; Möbius (1879–81), ii. 51.

23/14–15. This seems to be the last question in the commentary to *Háttatal*.

Stanza 24 is very similar to st. 47, except that the latter has identical syllables at the beginning and end of the odd lines, as well as at the end of the odd lines and beginning of the even lines. In Egill's *lausavísa* 4 (*ÍF* ii.

110), the first couplet is like st. 47, the second and last like st. 24. Cf. also *Háttalykill*, st. 33 (see JH–AH 127), Egill's *lausavísa* 16 (*ÍF* ii. 156), Hallfreðr's *lausavísa* 11 (*Hkr* i. 331).

Stanza 25. The first part of the even lines, i.e. up to the final disyllabic word, is parenthetical and in each case provides a gloss (*segir til*) to a kenning in the main statements in the odd lines (to which the last word in each of the even lines belongs). The device of providing a gloss to a kenning is used by Egill in *Hǫfuðlausn* 8/4. Stanzas 25–7 do not involve change of metre or arrangement of alliteration and hendings otherwise than to fix the position of the caesura in the even lines, and logically they belong with stt. 9–16, especially st. 13. All three stanzas involve regularly placed parentheses that occupy less than a full line. (Cf. Kuhn 1983, 167–8.)

Stanza 26. The main statements come in the odd lines and the first word (always a monosyllable) of the even lines. The remainder of the even lines are parenthetical and proverbial or gnomic statements (*orðskviðir*). Cf. st. 13.

Stanza 27. The parentheses in this stanza have no specified content, though they are positioned as in st. 26, with a caesura after the first syllable of the even lines (with a slight irregularity in l. 3; this also has a caesura after its first syllable, which belongs in sense to the second part of l. 2). The description in 27/9–11 could apply to st. 26 as well. Cf. JH–AH 132.

27/7. jarla prýði presumably refers to Hákon, the subject of this part of the poem, though according to *Sturl.* i. 278 it was Skúli who gave Snorri the ship mentioned in the next stanza.

27/13. með fullu háttaskipti: i.e. with variation in length of line or arrangement of alliteration or rhyme. *Þessi* (27/12) refers to st. 28, though in fact st. 24 too can be seen as involving variation in the hendings (cf. 23/14 and 1/44 n.; Möbius 1879–81, ii. 51–2); or alternatively as having linking of lines rather like the linking of stanzas in stt. 15–16.

Stanza 28: this is the metre of Hallar-Steinn's *Rekstefja* (*Skj* A i. 543–52), see Kuhn (1983), 333–4.

28/3–4: quoted in *TGT* 79 in illustration of *cacemphaton* (*-braks skylja*).

28/4. King Hákon may indeed have given Snorri a ship-levy estate (Snorri is said in *Sturl.* i. 278 to have been made a *lendr maðr*), but in view of the undoubted gift of a ship mentioned in the second half of the stanza, and the uncertainty about the ending of *skipreiðu* (see textual n.), it may be that the word should be *skipreiða* (*skipreiði* m. ship's gear or rigging).

28/6. The gift of a ship (though by Skúli, not the king) is mentioned in *Íslendinga saga* (*Sturl.* i. 278), cf. note to st. 95. *Dýr* is pl., though it seems unlikely that Snorri was given more than one. Because of the frequency of the generic sg. and the apparent occasional use of pl. when sg. is meant, as well as the similarity of some sg. and pl. forms, it is often difficult to know whether it is one ship or more that is referred to in *Háttatal*. See Glossary under *skip*, *sog*.

28/14. It is not clear whether it is permitted to have the first rhyme syllable delayed in the odd lines too; this only occurs in the even lines in st. 28. Cf. 35/3 and 7. Lines 28/12–13 perhaps mean that if the hending falls on the second syllable in the odd lines, it will still be *skjálfhenda* but no longer *tvískelft* (or maybe 'if the first hending falls on the syllable next after the first, the *skjálfhenda* [i.e. the position of the alliterating staves] will not be affected'). One of the characteristics of *skjálfhenda* seems to be that it uses type A lines with a 'heavy' first dip, which can carry the first hending; this is a common feature of Hallvarðr háreksblesi's *Knútsdrápa* and Hallar-Steinn's *Rekstefja* (see Sievers 1893, 107–8 and cf. 35/13 n).

29/1–2. The two visits were presumably in the winters of 1218 and 1219 (not Snorri's first and second visits to Norway), see *Sturl*. i. 271–2.

29/9–10. The fourth syllables of the even lines are all long stressed syllables (with the caesura before them) which form the first syllables of trisyllabic words concluding the lines. The even lines are consequently all Sievers type E, made up of two trisyllables. The name *detthent* ('falling rhymed') probably refers to this rhythm (´ ` ×), cf. JH–AH 131.

30/9–10. The third syllable in each of the even lines is a long stressed syllable, the first of a disyllabic word (which is involved in neither alliteration nor rhyme); the even lines are all made up of three disyllabic words with long first syllables and are consequently all Sievers type A.

Stanza 31 begins the (section of the) poem about Skúli. On this stanza see Kuhn (1969*a*), 228–30; (1983), 147. The odd lines are all type D with the first hending falling on the third syllable; in st. 32 it is the even lines that have this pattern.

Stanza 32: see Kuhn (1983), 165.

32/5. If *seim-* is read (as in TWU) the kenning would mean 'gold-spender'.

32/10–11: obscure. The rhymes in even lines normally involve syllables containing the same vowel (followed by the same consonant or consonant-group), so *lúkask á einum hljóðstaf báðar* may refer to the vowels in the following unstressed syllables, though in st. 32 they are only the same vowels in l. 2 (and in l. 6 in W, which reads *ilstafni*; even if *einum* were taken to mean 'a' rather than 'the same', the description would still only be applicable to l. 6). Similarly it would seem that *samhljóðandi sé eptir aðra* can only refer to the final consonants at the end of the unstressed syllables, though both rhyme-words have unstressed syllables ending in consonants in l. 4 and neither has them in l. 6. In a number of places it seems that the writer thinks of rhyme-words as normally having two syllables, cf. 38/10, 48/9–10, stt. 74–5 (which are distinguished by having monosyllabic words providing the first hending in each line; st. 45 also has monosyllabic rhyme-words), 76/11, 77/11–12, 78/10. Cf. also 36/10 (*lúkask allar einnig*), 71/11 (*endask báðar í einn staf*), 73/10 (*lúkask í einn staf báðar*), 78/10 (*lúkask báðar í einn staf*); the first and last of these phrases relate to monosyllabic rhyme-words, the other two to disyllabic rhyme-words where both syllables rhyme. Cf. *TGT* 50–2 and 98; the example there given of *riðhent* has two-syllable rhymes and illustrates *homoeoteleuton*. On two-syllable rhymes generally in skaldic verse see Kuhn 1983, 83, and note stt. 71–3; there are also examples in Kormakr's *lausavísur* 38/5 and 63/7 (with Bugge's emendation, see *Skj*. A i. 91), *ÍF* viii. 271, 301. Cf. 16/12–13 n.

33/2. TW read *reið*, R* has *riðu* (U *reidd*); with *sverð* as subject this verb would mean 'swing, be wielded'.

33/5 hilmis bróður: Skúli was (half-)brother of King Ingi Bárðarson (1204–17); stt. 32–7 refer to the events of 1213–14, while Ingi was still king. (That Ingi and Skúli had different mothers is implied by the use of the word *samfeðra* in *Hákonar saga*, ch. 9; but Skúli is in various places called Ingi's *bróðir*, and in the same chapter claims to be *skilfenginn*, 'legitimate'—cf. also ch. 88, where he describes himself as 'bróðir Inga konungs samfeðri ok skilgetinn'. Earl Hákon galinn, on the other hand, was Ingi's half-brother on the mother's side. See *Bǫglunga sǫgur* (1988), ii. 26–7.)

33/9–10. The writer identifies the antepenultimate syllable as the additional one in the even lines, and thus this syllable never takes part in the alliteration or rhyme. Preceding this syllable each of his even lines is Sievers type A2k (first dip a long, half-stressed syllable, second lift a short syllable by licence, Sievers 1893, § 61.4) or perhaps D2 (since the second syllable carries the rhyme and may be fully stressed); but if the third syllable in each line were considered to have resolution the fifth syllable would constitute the second dip and would not then be 'extra' (type A2ab or D1). Cf. st. 8, where resolution provides at least one 'extra' syllable in seven out of the eight lines and see Kuhn (1983), 68.

34/9–10. Here it is the fourth syllable from the end (or the fourth from the beginning) that is identified as the additional one; it could be regarded as providing resolution with the preceding short syllable, which if the lines are taken to be Sievers type D would be half-stressed, or the fifth syllable in the line could be regarded as an extra unstressed syllable in the dip of a type D line, with the half-stressed syllable short by licence (D2, Sievers 1893, § 61.4). The requirement that the first word in the even lines be a superlative adjective introduces a grammatical and non-metrical element into the variation. Cf. st. 9 in *Víglundar saga*, *ÍF* xiv. 104, l. 2.

35/11. Þorvaldr veili's only recorded stanza (see *ÍF* xii. 262) does not use *skjálfhenda*. The poem referred to at 35/12–13 does not survive. There seems to be an implication that the name of the poem (and the metre) was connected with the shivering produced by the cold, rather than from the sound of the verse-form, though it seems possible that the name originally referred to the sound of alliterating staves coming close together early in the line (see next note; the name *tvískelft*, st. 28, would support this: that stanza had *skjálfhenda* twice in each half-stanza), in which case Snorri's note is based on folk-etymology (or Veili's poem had nothing to do with the metre *skjálfhenda*). The term *kviða* is often used of poems that have no refrain, being found in the names of many narrative poems as well as of poems in the metre *kviðuháttr* (see Appendix and Wessén 1915, 129), but the phrase *drápan steflausa* seems a contradiction in terms, since a refrain is usually considered an essential feature of a *drápa* (though indeed *Íslendingadrápa*, *Skj* B i. 539–45, is without one).

35/13. It seems clear from Snorri's three accounts of it that for him the essential feature of *skjálfhenda* was the occurrence of the alliterating staves in the odd lines on the first and third syllables, rather than any particular arrangement of the hendings, though in 35/3, 7 this arrangement of the *stuðlar* is combined with the placing of the first hending on the second syllable of the line, as in *Háttalykill*, st. 41/3 and 7, which Kuhn (1983), 105, 289, 333–4 takes to be an essential feature of the pattern; in *tvískelft* (st. 28) this arrangement of the alliteration is combined with the first hending falling on the first syllable of the line, in *in forna skjálfhenda* (st. 35) there are *skjálfhendur* in the third line of each half-stanza only, combined with *aðalhending*. Lines 39/9–11 imply that *in nýja skjálfhenda* had *skjálfhendur* in the odd lines combined with *skothending* (as in *Háttalykill*, st. 41). All *skjálfhent* lines are characterized by having a 'heavy' first dip (Sievers type A2a), whether or not the first hending falls upon it.

Stanza 36. On *þríhent* see JH–AH 130, where parallels in Latin verse are quoted.

36/7–8. In *SnE* (1952), 471 *egg- . . . -þing* is taken as a compound separated by tmesis; this figure is also assumed at 19/3–4 (*grunnfǫll*; cf. *grunn* in Glossary) and 62/6–7 (*folkskúrum*) by Finnur Jónsson in *Skj* B ii. 66, 78. But although tmesis undoubtedly occurs in skaldic verse—the clearest examples involve the splitting of compound proper names, but there are cases with other nominal compounds too, especially kenning compounds; see Finnur Jónsson (1933); Reichardt (1969); Amory (1979); Kuhn (1983), 108–9, 111–12—all three supposed examples in *Háttatal* are capable of more satisfactory interpretations. See also 15/1–2 n.

36/10. lúkask allar einnig: sc. all the rhyme-words are monosyllables (or monosyllabic second halves of compound words; they have no ending syllables); *fylgir samstǫfun fyrir hverja*: each is preceded by a (non-rhyming) syllable. According to normal scansion the rhyming syllables in the even lines of this stanza are all in the dips (half-stressed if taken as type A2ab), and normally *dróttkvætt* lines end in a disyllabic word with the second syllable unstressed.

Stanza 37: On *hinn dýri háttr* see JH–AH 129.

37/5–8. E. A. Kock (*NN* 1311) tries to simplify the somewhat tortuous syntax of this half-stanza by taking *hátt* as an adverb ('nobly') parallel to *þrátt* (instead of as an accusative noun, object of *bjóða*) and linking the whole of 37/5 with the following line ('the prince of the clan of *hǫlðar* distributed red wealth nobly, repeatedly'). He then takes *þat* as the object of *bjóða* and *þjóð* as the subject of *stóð*, and is forced to accept the reading *of* (R*W; U has *um*) in place of RT's *af*. This is undeniably smoother, but the use of *hátt* is unusual, *jǫfurr*, though frequently used in the poem, does not elsewhere appear with dependent genitive, and the statement 'people stood around the ruler' fits ill into the context.

37/9. í fyrsta ok þriðja: in the even lines too.

38/10: i.e. an unstressed syllable follows each rhyming syllable, in contrast to st. 37, where the first two rhyming syllables in each line were monosyllables (cf. st. 36). The first two rhyme-syllables in each line of this stanza, moreover, are all short, and with the following unstressed syllables provide resolution (giving Sievers type A2a or D lines; type D4 or E in ll. 4, 6, and 8). It is unusual for the penultimate syllable of a line also to be short as in l. 6, but this may be the result of the difficulty of finding a long-syllable rhyme for the short syllables earlier in the line, and short penultimate syllables are used in st. 90. See notes on stt. 8 and 16/12–13, and cf. Kuhn (1977); (1983), 80–2. Short and long consonants in pairs of rhyming syllables, however, as in ll. 2, 3, 5, and 7, are not all that uncommon (Kuhn 1983, 77–8).

39/5–8. King Ingi (*hilmir foldar*) invested Skúli with an earldom in 1217.

39/9–10. It could be the echo of *hjaldreifan* (l. 1) in *hjaldrs* (l. 5) that is being referred to—or possibly that of *hjálmar* (l. 4) in *hilmir* (l. 5), cf. stt. 16 and 24—but by analogy with st. 15, also described as *tiltekit* (see textual note on 15/1), it would seem to be the syntactical linking of the two halves of the stanza by *þá er* that is the feature meant. Stanzas 39 and 40 would then both be syntactical rather than metrical variants and would belong logically with stt. 9–16 and 25–7. It is presumably accidental, rather than an essential part of *tiltekit*, that the alliteration in these lines (*hjálmar* | *hjaldrs þá er* [or *þás*?] *hilmir*) is reminiscent of *skjálfhenda* (39/10–11). The writer does not explain how *skjálfhenda in nýja* differs from *in forna* (st. 35, cf. st. 28), but it is presumably that in the former the lines that have *skjálfhendur* have *skothending* rather than *aðalhending*.

40/1–4: quoted in *TGT* 94–5 as an example of anaphora. The stanza is also similar to *áttmælt* (st. 10). Cf. *FoGT* 150–1 and Eysteinn Ásgrímsson, *Lilja* 62 (*Skj* B ii. 406), quoted by P. G. Foote (1982), 114; (1984), 256–7;

also JH–AH 125–6 (who quote *PE Helgakviða Hundingsbana* II 5–6 and *Skj* B ii. 157 st. 48, *Sturl.* i. 428); the quatrains in *Grettis saga*, *ÍF* vii. 151–2 form a similar stanza too. See Vésteinn Ólason (1969).

40/7–8: quoted in the additions to *Skáldskaparmál* in W in illustration of descriptions of men as distributors of gold (*SnE* 1924, 105).

40/9. með orðum could mean either 'with its lines' or perhaps more likely 'with words', i.e. the same as *með máli* (1/47–8, 8/45, 9/11): 'varied as regards *dróttkvætt* in meaning'—but not in sound. In any case this is another syntactical variant: it is identical with *dróttkvætt* metrically (*breytt til*: 'based on'?), and does not conform to the description at 23/14–15. Note, however, that the repetition of *hverr* and *hann* as internally rhyming words in ll. 1–4 and 6–8 has resulted in virtual end-rhyme in these lines too (cf. st. 99 and n.; also stt. 41 and 53).

41/9: i.e. the alliteration falls on the same syllables as the rhymes. Cf. stt. 53 and 58 (and also stt. 24, 46, 47).

42/9. Cf. 8/18.

42/12. í inu fyrsta vísuorði: and in the third, too; and the *skothendur* are double in each case.

Stanza 43, *in minni alhenda*, probably ought to come after st. 44, *alhent*, as it does in W.

43/9–11. The author here shows his concern for consistency throughout individual stanzas; cf. 51/9–11, 53/11–13, 58/14–16; also 1/26–7, 4/18–20 and n., 8/15–17.

43/10–11: i.e if one or more *skothendingar* occur in an *alhenda* stanza.

Stanza 44. On *alhent* see Kuhn (1983), 305–6.

44/7. E. A. Kock (*NN* 3146) points out that the preposition *við* should not really carry both alliteration and rhyme, and suggests that it should be taken as postposition (thus capable of carrying a primary stress) with *herfjǫlð*, and that *valdi* should be taken as nominative and subject of *friðask*. There are other places, however, where the metre forces the poet to stress unusual words (e.g. *hann* 1/2, 7/3, 8/7, *fram* 20/4, *fyrir* 80/2, *hans* 80/5, 89/6, *þar* 88/8, *honum* 93/2).

44/15–18. Not recorded elsewhere (*Skj* A i. 534, B i. 515). The metre would be more regular if contracted forms were used (*bragarmál*), i.e. *baðk* in l. 1 and *gǫrs* in l. 2. The first lift of l. 3 has resolution; cf. st. 8. On the line-type in ll. 1 and 3 see Kuhn (1983), 156–9; it also appears in 44/1 and 7.

Stanza 45. Cf. Kuhn (1969*a*), 226–7; (1983), 157, 289.

45/11–12. sem riðhendur: i.e. together at the end of the line, cf. stt. 32, 55, 56.

Stanza 46. *Samhent* is rather like *liðhendur*, st. 41, except that its odd lines have *aðalhending* like *rétthent*, st. 42, and the alliteration does not invariably fall on the same syllables as the rhyme in even lines. In addition, *liðhendur* has *skothending* between odd and even lines; *iðurmælt* (st. 47) has *aðalhending* between them.

46/10. með einum stǫfum: i.e. they have the same initial sounds (as well as final ones).

Stanza 47 (*iðurmælt*) corresponds to *Háttalykill*, st. 29 (though this is less regular than Snorri's stanza); cf. JH–AH 127. The closest parallels are perhaps Egill Skallagrímsson's *lausavísa* 4/1–2 (*ÍF* ii. 110) and Hallfreðr's *lausavísa* 11/1–2 (*Hkr* i. 331).

47/10–11: i.e. st. 24 (see textual note on 24/1).

Stanza 48 (*klifat*) is similar to *Háttalykill*, st. 12; see JH–AH 127.

48/9–10. taka með aðalhending ina síðari: i.e. the *aðalhending* in the odd lines rhymes also with the *aðalhending* in the even lines (as indeed they also do in *iðurmælt*, st. 47); in addition there is *skothending* between the rhymes in the first half-stanza and those of the second. On the device of having the same word repeated in several successive lines cf. Hallfreðr's *lausavísa* 11 (*Hkr* i. 331).

Stanza 49 contains a series of allegorical references to Hildr Hǫgnadóttir, whose name means 'battle'. Hildr is also the name of a valkyrie (e.g. *PE Vǫluspá* 30) and is frequently used in kennings for battle and armour; both persons can be used as personifications of battle and they are not always kept clearly distinct. The device is a form of *ofljóst*.

49/9. it fjórða vísuorð: i.e. of each half-stanza, as usual; similarly 50/9.

49/10. skal setja með hending: i.e. is included after the second rhyme-syllable. Cf. 32/10–11 n.

51/10. annarr helmingr stýfðr: i.e. in one line only, as contrasted with *tvístýfðr* and *alstýfðr*.

Stanza 53 differs from st. 41 in that instead of *aðalhending* in the even lines, which forms *skothending* with both rhymes in the odd lines, there is *skothending* in both odd and even lines, but *aðalhending* between the first rhyme-syllables in each pair of lines. (On rhymes such as *-ór/-jór* see Kuhn 1983, 76.) Part of the sentence describing this is omitted in all manuscripts; it must have continued in some such way as '. . . there are full rhymes in the second and fourth, and on the other such that with the second rhyme-syllable there is half-rhyme in the second and fourth lines' (*. . . eru aðalhendingar í ǫðru ok fjórða, en aðrar at við ina síðari hending eru skothendingar í ǫðru ok fjórða*). There is the further difference between stt. 53 and 41, which is not pointed out in the commentary as it stands, that st. 53 has *oddhending* in all lines, while st. 41 has it only in even lines. See *NN* 2182; Sievers (1893), 108.

53/11–14. Here again (as also at 51/9–11) the writer acknowledges that he is making a systematic use (*eru nú settir saman*) of what in earlier poets had been sporadic lapses or variations of metre found in occasional lines rather than in every line of a stanza. Cf. 58/14–16, where his tendency to prescriptiveness in this matter is most marked.

53/14. None of the verses attributed to Ragnarr in *Ragnars saga loðbrókar* seem to be consistently in this metre, though some of them, like *Krákumál* too (*Skj* B i. 649–56; no longer attributed to Ragnarr), have occasional *aðalhending* in even lines and frequently no hendings in the odd lines; but it is doubtful whether any of them are genuine anyway. The features of st. 54 are actually found more commonly in verse attributed to Bragi and Egill.

54/10–11. Cf. 1/11–13. The inclusion of an unstressed syllable before the *hǫfuðstafr* is a common feature of early skaldic verse (e.g. that of Bragi and Egill Skallagrímsson) as well as of verse attributed to Ragnarr, including *Krákumál*; see Kuhn (1983), 168–9.

Stanza 55. Of the five *lausavísur* attributed to Torf-Einarr in *Skj* A i. 31–2, none corresponds exactly to Snorri's example, though the odd lines nearly all lack rhyme and the even lines frequently have *skothending*. But some couplets have *skothending* and *riðhendur* in the even lines (1/1–2, 5–6, 2/7–8, 3/5–6, 4/3–4, 5/5–6). Line-lengths could be made more regular by reading *sé* in l. 1, *eðr gjarnara at* in l. 3, *þás* in l. 7.

Stanza 56. The features of *Egils háttr* are found in occasional couplets in Egill's *lausavísur*, never in a whole stanza; see his *lausavísur* 6/1–2, 19/5–6, 40/7–8 (*ÍF* ii. 119, 163, 269). The odd lines of his stanzas frequently lack hendings, however.

56/7. All other manuscripts, and R* as well, have *hann* for *herr*; but both readings make good sense.

56/8. There seems to be no advantage in reading *framkló loðnar* (or *loðna*: cf. *Skj* B ii. 76); see Fidjestøl (1982), 72–3. *Kloð-* in *Skj* A ii. 67 is an error.

57/9–10. On the rule for the position of the second hending in the line see Kuhn (1983), 89–90. Lines with the second hending not on the final stressed syllable are another feature of the earliest skaldic verse.

Stanza 58: see Kuhn (1981). None of Bragi's surviving stanzas (*Skj* A i. 1–5, B i. 1–5) has all the features Snorri specifies for *Braga háttr*, though in a number of places the lack of hendings in one line of a pair is so to speak compensated by a rhyme between one syllable in that line and the hendings in the other line of the pair; but this only rarely seems to form *aðalhending* with either of them. See *Ragnarsdrápa* 4/5–6, 7–8, 9/5–6, 7–8, 13/3–4, 16/3–4, Bragi 2. 3/1–2, 3–4. *Skothending* between as well as within pairs of lines is found in *Ragnarsdrápa* 11/3–4, 7–8, 18/3–4; cf. also *Ragnarsdrápa* 17/2–3 and Egill's *lausavísa* 27, *ÍF* ii. 200. In none of these cases are the hendings positioned as in *Fleins háttr*, though the two features are found (in different lines) in the stanza in *ÍF* ii. 224–5 (cf. 57/9–10 n. and see Kuhn 1981). In ll. 3–4 of Snorri's example, his *aðalhending* appears to be formed between syllables with long and short vowels (*ógn-* ~ *togn-*), but see Noreen (1923), § 127. 3. Cf. Kuhn (1983), 80–2.

58/14–16. Inconsistency in the arrangement of hendings is especially characteristic of Bragi's verse, and of that of many other very early skaldic poets. Later poets became more regular (see Kuhn 1981). On the use of *háttr* to mean primarily the pattern of hendings being used see 0/1 n.

58/17. On *kimblabǫnd* see JH–AH 129–30. The device seems not to be a traditional Norse one, and is probably adapted from French and Latin verse.

60/3–4. hræs seems to add a redundant gen.; *hræs hrannir* would be a satisfactory kenning for blood, as would *hrannir fella* (cf. 15/1–2 n.). 'Waves of corpse-fellers' would, however, also be a possible kenning for blood, and there seems no reason to reject the readings of T, W, and (probably) R. The correction that Finnur Jónsson claimed (*SnE* 1931, 240) had been made, to *hvatt*, cannot now be read, and though it would perhaps give easier sense, is of doubtful authenticity (it is read *hraustr* in *Skj* A ii. 68). Kock in *NN* 2184 proposed taking *fella* as inf. parallel to *gella*, though this is stylistically unlikely, and in any case *falla* would be the expected verb.

60/8. It has been questioned whether *fólk* ever really does mean 'sword' in Old Norse poetry; the sense-development is certainly problematical (see Richardson 1975). It is listed in the *þula* of sword-names (or of names of parts of swords) in *SnE* (1931), 202, and though the compiler may have included it as a result of misunderstanding *fólk* (*er*) *í dreyra* in Gunnlaugr Leifsson's *Merlínússpá* II 66 (*Skj* B ii. 37), where the word perhaps really means 'warriors', Snorri could easily have made the same assumption, and is likely also to have known the *þula*. There does not, therefore, seem to be any reason why Snorri should not have intended the meaning 'sword' here and possibly also in *fólkskúrum* (62/6–7; see next note), even if the meaning was not in fact sanctioned by earlier poets. The expression *lætr sund unda fylla fólk*, however, is odd whether *fólk* is taken to mean 'battle' or 'sword'. Kock (*NN* 2184) proposed taking *fólk* as the first half of a compound with *sund*, meaning 'mighty sea' (obj. of *fylla*; cf. *her-*, *þjóð-*); *stinn* in l. 6 would then either have to be emended to *stinnr* and taken with *stiklir* (so *Skj* B ii. 77) or, as suggested by Kock, taken as first half of a compound *stinnmen* 'stiff neck-ring'. A further possibility would be to read *lætr stinn fólk fylla sund unda* 'lets firm warriors fill a sea of wounds.'

62/5–8. Kock (*NN* 191) suggested that *vísi* should be taken as subject of *heldr* rather than of **skiptir* (which necessitates adopting the reading of W for this word and taking *heldr* as impersonal), *fólk skipta svá* as a parenthetical clause (*fólk* cannot, however be taken as object of **skiptir*, since this verb normally takes the dative), *boga driptum* as object of *heldr* rather than of **skiptir*, *skúrum* as adverbial dative rather than as object of *lýstr*, and *Mistar regni* as object of *lýstr* rather than of *heldr*. This has the advantage both of avoiding the emendation of *skipta* (if *fólk* is taken as subject) and the highly doubtful compound *fólkskúrum*, which involves both the questionable meaning of *fólk* = 'sword' and the use of tmesis. Cf. 36/7–8 n., 60/8 n. Kock's arrangement also results in symmetrical organization of the parenthetical clauses in each couplet of the stanza, rather than the tortuous syntax of *Skj* B ii. 78.

62/9–12: i.e. in this form of *hrynjandi* the first two syllables in the odd lines and the fifth and sixth in the even lines are not involved in the alliteration or rhymes (the *stuðlar* come near the end of the odd lines); Snorri always identifies the 'extra' syllables in verse-forms that have more than six in each line in this way (cf. 33/9–10 and n.; when he says *þá er eptir sem dróttkvætt*, he means, of course, that the rhythm would then be as *dróttkvætt*, though the sense would be destroyed). The rhythm in the odd lines tends to be trochaic, i.e. the four syllables after the first two are Sievers type A, while in the even lines the first four syllables constitute Sievers types D or E. There are no resolutions, but note that the fifth syllable in *hrynjandi* is often short (62/6, 63/8, 64/6, 7).

Stanzas 63–6 refer to events of 1221, when Skúli defeated the Ribbungar in the Vík and brought about the death of one of their leaders, Gunnarr Ásuson. Stanzas 63–4 and 66 are quoted in connection with the account of these events in *Hákonar saga Hákonarsonar*, chs. 74–5.

63/9–11: i.e. in every line the first hending falls on the third syllable (the first of a two-syllable word) and is followed by three unrhymed syllables (though they are not all unstressed) before the second hending; the rhythm is entirely trochaic (all Sievers type A lines). The 'extra' syllables in this stanza would presumably be identified as the fifth and sixth in the even lines and the first and second in the odd lines (although the first syllables carry the alliteration). The occurrence of alliteration and the alternation of *oddhending* and *aðalhending* are as in *hrynhenda* (= *hrynjandi*, st. 62), though their positions within the lines are not identical; in st. 62 the first rhyme-syllables were the third syllables of the line in odd lines but the second in even lines.

63/12. *einn* perhaps ought to read *enn*; the sentence is lacking in W.

64/9. óbreytt: this stanza (64) differs from st. 62 in that the *stuðlar* fall on the first and third (instead of the third and fourth) lifts of the odd lines (as in st. 63), and from st. 63 in that in three out of the four even lines the first rhyme (which falls on the second lift, or on the half-stressed element of a type E line) is on the second syllable of the line (as in st. 62), instead of the third (as in the odd lines and every line of st. 63). The odd lines have trochaic rhythm, and so has l. 4 (*fyrir* probably to be pronounced *fyr*), but ll. 2, 6, and 8 begin with Sievers types D or E. This stanza thus has its odd lines like 63 (except that the caesura falls after the third instead of the fourth syllable) and its even lines (predominantly) like 62. Cf. Möbius (1879–81), ii. 47. Lifts fall on short syllables in 62/6, 63/8, 64/6, 7. Except in 63/8 (and perhaps 64/7), these are under the regular conditions for licence (i.e. there is an immediately preceding long stressed syllable, see Sievers 1893, §§ 9.2 and 61.4). *Serks* in 64/4 does not alliterate, cf. st. 76 n.

65/9–14: i.e. each line begins with Sievers type A*2 or D*; the fourth syllables of ll. 4 and 6 are short, i.e. these lines are type D*2 (cf. note to 62/9–12). The metre could be regarded as *málaháttr* (or *Haðar lag*, st. 79) with an added trochee, just as *dróttkvætt* is usually regarded as *fornyrðislag* with an added trochee and *hrynjandi* as *dróttkvætt* with an added trochee. Cf. Sievers (1893), 112, 240.

65/14. eigi mjúkt: i.e. this metre sounds jerky (a heavy caesura tends to come after the third syllable); it presumably does not mean it is no longer *mjúkt* if the syllables are taken away. This metre seems not to be used except in *Háttatal* and *Háttalykill*, st. 4.

Stanza 66. Cf. st. 55; the only difference is in the position of the *frumhending*.

67/2. This clause is dependent on *siklinga*.

67/9–10. sem í dróttkvæðum hætti: except that the *hǫfuðstafr* is not at the beginning of the even lines (cf. 1/11–13), as is also permitted in st. 54; in both stanzas the even lines are all types B or C. 67/8 appears to have double vowel alliteration (*oss*, *jǫfrar*).

67/10–12. Stanzas 1–8 are all counted as one verse-form (regular *dróttkvætt*), since the variations were only rhetorical or examples of *leyfi*; cf. 1/52, 6/17–18, 8/45–50. Stanza 9 is counted as the first verse-form that is distinct from *dróttkvætt*. At 67/11 *fyrst* means at the beginning of *Háttatal*.

67/14. it þriðja kvæði: see 69/4–5 n.

Stanza 68: all lines are regular Sievers types whether or not the *ek* is contracted in ll. 3, 5, 7, so that as far as the metre is concerned this stanza is *fornyrðislag* (like stt. 69–71). There are five syllables in l. 2, but the first two constitute resolution, so they are the equivalent of four. The final stressed (half-stressed?) syllables in ll. 2, 6, and 8 are short by licence; they are type A2k—though some of these lines could perhaps be taken as type D, like l. 4, cf. Sievers (1893), 33, 100, 112. Note the *aðalhending* in syllables apparently containing *æ* and *œ* in l. 4 (cf. *LP* s.v. *mærð*; but it is doubtful if there are enough early examples of *mær-* rhyming with *-œr-* to confirm the existence of an alternative form *mœrð*). Line 1 lacks both a fifth syllable and *skothending*, presumably because it belongs with 70/8 in sense (and imitates the *stef* 'Knútr vas und himnum' in Sighvatr's *Knútsdrápa* (*Skj* B i. 232–4), which also lacks *skothending* though most other lines in the poem have hendings; see JH–AH 60). Stanza 68 has *skothending* and a single *stuðill* in the odd lines (except l. 1) and *aðalhending* in the even lines, st. 69 has two *stuðlar* but no hendings in the odd lines and *aðalhending* in the even lines, st. 70 has *skothending* and two *stuðlar* in the odd lines and *aðalhending* in the even lines. All three variations are found in *tøglag* poems (see Appendix), but only Einarr Skúlason's *Haraldsdrápa* II (*Skj* B i. 425–6) seems consistently to use one of them (it follows the pattern of st. 70).

68/11. en: perhaps this should be normalized *in* (cf. 69/5).

Stanza 69: there is again resolution in ll. 2 (type D4 or E?) and 8 (type D2). Line 7 has only four syllables (type A2k). Cf. Sievers (1893), 112.

69/4–5. þrenn . . . fjórða: cf. 95/3. Snorri seems to be referring to the second and third sections of *Háttatal* (cf. 30/11 and 67/14) and two other poems about Skúli, making four in all. Cf. Finnur Jónsson 1920–4, ii. 78; *Sturlunga saga* i. 278; Möbius (1879–81), i. 34. See *þrennr* in Glossary.

69/12. hinn þriði háttr: i.e. the third kind of *tøglag*.

Stanza 70. Like st. 68, with both *aðalhendingar* and *skothendingar*, but with two *stuðlar* rather than one.

70/1–4: is Snorri claiming to be using new metres, previously not employed by court poets? Note that *mœti* is indefinite; it does not seem to mean just that such poetry has never before been composed about Skúli.

70/10–12. Cf. st. 8. As there, some of the extra syllables in stt. 68–70 are the result of resolution (*skammar samstǫfur*?), some are unstressed enclitics which can usually be contracted (*skjótar samstǫfur*?—but at 83/10 *skjótar* seems to refer to resolution too, at 86/10 to both resolution and enclitics). In st. 70 the first is the only line with five syllables.

70/12–16. Snorri provides as an example of a *stef* 68/1 and 70/8, which together make up a sentence independent of the rest of the three stanzas (i.e. it is a *klofastef*). Stanzas 68–70 thus exemplify one *stefjamél*, and a complete *tøglag* poem would contain several such sections, with the beginning and end of each marked by the occurrence of the lines of the *stef*. Snorri does not indicate whether each *stefjamél* can be marked by a different *stef*, as in Egill's *Hǫfuðlausn* (*Skj* B i. 30–3)—though this poem is not in *tøglag*.

Stanza 71 differs from st. 70 in that in it the *aðalhendingar* are two-syllable rhymes, as in stt. 72 and 73. Cf. 32/10–11 and n.; see also JH–AH 129 and Kuhn (1983), 83.

72/9–10. But in st. 72 there is only one *stuðill* in ll. 3 and 7.

Skemri orðtǫkin: the stressed syllables in ll. 4, 6 and 8 are short; in l. 2 they end in vowels (the result of loss of -*w*-), and Snorri may have considered these also to be short, cf. Hreinn Benediktsson (1968); Sievers (1893), 58; Noreen (1923), § 49; Heusler (1950), 15; Kuhn (1983), 54. Sievers (1893), 113 compares these light lines to occasional occurrences of somewhat similar lines in *tøglag*. The odd lines are regular

Sievers types, though l. 7 is type C with delayed alliteration; see Kuhn (1983), 50; Sievers (1893), 38.

Stanza 73. All type A lines and two-syllable rhymes (cf. st. 71 n.). This verse-form is quite possibly Snorri's invention, cf. 70/1–4 n.

73/1–4: quoted in *TGT* 98 in illustration of *homoeoteleuton*.

73/2: normalized *Rǫkkva dǫkkva* in *Skj* B ii. 80; but there is insufficient evidence for the existence of an alternative form of *døkkr* with ǫ (see *LP* s.v. *døkkr*; Noreen 1923, § 167), and the vowel in the sea-king's name can hardly be said to be certain even though it rhymes with *Hlǫkk* in a stanza of Bjǫrn Hítdœlakappi (*ÍF* iii. 140). In any case by Snorri's time the vowels *ø* and *ǫ* would probably have been sufficiently similar for them to be used to provide *aðalhending* together (cf. note on st. 68). R's *støkkva* cannot be right.

73/10. lúkask í einn staf báðar: cf. 71/11 and 32/10–11 n.

73/11. In view of 74/11 and 75/11, Möbius (1879–81), ii. 103 assumed that the names of the verse-forms in stt. 74 and 75 were accidentally reversed in R (they are not given in any other manuscript; the verse-form of *Háttalykill* that corresponds to *Háttatal*, st. 75 is there named *háhent* (st. 15), cf. JH–AH 64).

Stanza 74. Since the first hending falls on the second syllable, all lines must be type A2al (secondary stress on first dip); similarly the even lines in st. 75.

75/4. It is awkward to have to take *frár* with *fleinþollr* and *mál* with *falla*; in the other even lines of this stanza there is a compound word in this position, and one would at least expect that *frár* and *mál* would go together; but Kock's suggested *fœr* (*NN* 2185), even though it can be justified because the original reading of R is unclear (W has *fjǫr*), cannot really go with *mál*. The adjective means 'passable' (of a road) or 'capable' (of a person), though it is found also in the phrase *fœrt veðr* ('weather in which it is possible to travel'; e.g. *Sturl.* ii. 144).

75/9–10. Lines 3 and 7 seem to be type A2b with anacrusis, an unusual feature to find in skaldic verse. Line 5 is A2ab with resolution of the first stress, l. 1 A*2b. Line 3 has six syllables unless *þá er* is read *þás*. The even lines are all type A2a. See Sievers (1893), 113.

75/11. stýfð: i.e. monosyllabic.

Stanza 76. Lines 1, 3, and 5 are not Sievers types; their rhythm is like that of *hrynjandi* (stt. 62–4) with the final syllable omitted (cf. st. 91). The third word in each of these lines has a short syllable followed by an ending syllable, so they could be taken as type D* with an added long syllable at the end. Line 7 has only six syllables and as a whole (i.e. including the final syllable) can be taken as type A2b; Rask (*SnE* 1818, 262) and later editors have added *fram* after *hlunna* to make it the same as ll. 1, 3, and 5. Line 6 seems to have double alliteration, but whereas *sv-* and *sn-*

were probably considered to alliterate together and with *s-*, *st-* usually alliterates only with *st-* (like *sk-*; cf. stt. 64, 73, 83, 91 and see Kuhn 1983, 50). Line 8 has *aðalhending* and l. 1 might be considered to have *skothending*: Finnur Jónsson (*SnE* 1931, 246 and *Skj* B ii. 82) emends *bekk* to *borð*, following W; the sense is not affected, but it avoids the possible *skothending* with *strýk-* which is not required.

76/11. stýfð: i.e. the first rhymes fall on monosyllables or long stressed syllables (the first parts of compounds) followed by other long and probably half-stressed syllables, rather than ending syllables. Cf. 32/10–11 n.

Stanza 77. Cf. Ormr Steinþórsson 1, *Skj* B i. 385–6 (unless the poem is *stýft*, st. 51) and Óttarr svarti, *Óláfsdrápa sænska* (*Skj* B i. 267). See Kock (1933).

77/9–10. If all the main verbs are taken to be unstressed, each line is type A2b with a long, presumably half-stressed syllable in the second dip (l. 4 type A2ab? or D*4?), though ll. 3, 5, and 7 can also be taken as type A*2. There are resolutions in ll. 2, 6, and 7, and perhaps l. 4, but there is no way of making each line equivalent in number of syllables. Lines 1 and 8 are irreducibly six syllables, ll. 3, 4, and 5 have only five. No lines have seven syllables.

77/12. rétt í dróttkvæðu: i.e. the hending syllable is followed by an unstressed syllable, as is normal (though by no means invariable) in *dróttkvætt*. Cf. 32/10–11 n.

Stanza 78. All lines type A2ab, *stuðlar* always on the first and third syllables, hendings on second and fourth. Cf. JH–AH 130.

78/10. báðar í einn staf: perhaps refers to the fact that all hendings are monosyllabic (i.e. are not followed by an ending syllable; this phrase would then be tautologous with *ok allar hendingar hneptar*), or perhaps to the fact that the whole consonant groups at the end of the hending syllables are identical, unlike e.g. *lauki* ~ *raukn* 77/6, *heldr* ~ *skjǫldum* 24/7, *spjǫr* ~ *ǫrvar* 9/8. Cf. 73/10, 71/11, 36/10, etc.

Stanza 79: All lines type D*, all hendings on first and penultimate syllables, all *stuðlar* on first and third syllables. Metrically, the stanza is *málaháttr*.

79/6 Finnur Jónsson (*SnE* 1931, 247; *Skj* B ii. 83) adopts the reading of R*W, but while *falbroddr* may be a more normal kind of expression in skaldic verse, *valbroddr* is not impossible.

79/10. sem í dróttkvæðum hætti: but l. 3 has *aðalhending* instead of *skothending*. In l. 7 the rhyme syllables have long and short vowels (cf. Kuhn 1977; 1983, 80–2; 38/10, st. 58 nn.).

79/12–13. í sínu vísuorði hvár hending: i.e. the two rhyme-words are in successive lines, instead of in the same line as in *dróttkvætt*. In *full* (or *rétt*) *runhending* the same rhyme is used throughout the stanza.

80/6 has only 3 syllables; R* adds *til* before *hróðrs*. Otherwise the stanza (like st. 81) is metrically similar to *fornyrðislag* (or its variants in stt. 97–8) though the lifts come on some unusual words (*fyrir* l. 2, *hans* l. 5).

80/10–11. Því er þetta runhent kallat: the reason, however, is not clear. If the name refers to the 'run' or series of rhymes running through the stanza, the versions in which the rhyme changes within the stanza would not be the reason for the name. Cf. *runhenda* in Glossary.

81/10–11 must refer to st. 82, in which the lines are one or more syllables shorter than those in st. 81 and the rhymes are masculine ones rather than feminine as in stt. 80–1. Cf. st. 102, where the odd lines have three syllables.

82/9–10. tvau vísuorð sér um hending: i.e. each pair of lines has a separate rhyme.

82/10–12. hljóðstafr does not seem ever to refer to alliterating staves elsewhere in Old Icelandic (see Glossary), though (like *ljóðstafir*) it does in modern Icelandic (see Helgi Sigurðsson 1891, 26). If it does do so here, it may mean that single alliteration in the odd lines is permitted in this metre (*einu sinni fyrir málsorð*: '[only] once at the beginning of a word'; cf. 85/10 and see JH–AH 49). It is possible, however, that the sentence refers to the fact that the rhyming vowel in l. 2 constitutes a complete word (*á*: 'the vowel that constitutes the rhyme stands on one occasion in place of a word'). The phrase *ráða kveðandi* elsewhere refers to alliteration, however (1/13, 54/10), and the text of W unequivocally makes *hljóðstafr* refer to alliteration by adding *í runhendum háttum at í fyrsta ok þriðja vísuorði* after *rangt* in place of *ef*.

Stanza 83: all lines seem to be type D*1 (or C*1)—though l. 1 has two short syllables (resolution?) after the first lift—except for l. 6, which is type A with anacrusis. The sixth syllable in l. 8 is caused by resolution of the second stress (83/10 *ef skjótar eru*: cf. 70/10–12 and n.). On the alliteration in l. 8 cf. note on st. 76. The rhythm of the stanza is like *málaháttr* (see st. 95). Cf. Sievers (1893), 115.

83/2. eyðir as it stands must be vocative (addressing Skúli). The vocative, however, is, usually accompanied by an imperative elsewhere in the poem (e.g. 3/7–8, 30/5–8). Finnur Jónsson (following Rask's suggestion, *SnE* 1818, 264 n. 1) in *Skj* B ii. 84 emends to *eyðis* (dependent on *hestar*; cf. variants in *Skj* A ii. 73).

83/5–6: quoted in *TGT* 52 in connection with a discussion of hendings; cf. 32/10–11 n.

83/9. ort: for the lack of concord cf. 41/10 (though there the word-order makes it less striking). *Með fullri runhending*: i.e. there is the same rhyme throughout the stanza.

84/9: i.e. the rhymes are masculine ones. The *stuðlar* also come in adjacent syllables at the end of the line (cf. st. 45). The stanza is not easy to scan;

all lines have four syllables and could be type A with long, half-stressed syllables in the second dip, but then the *hǫfuðstafr* in l. 2 would be in the wrong place and there are other irregularities. If one thinks of the final dip in each line as having been omitted (*hneptr*), they could all be either type C or D*.

85/10. ok fylgir þeim einn hljóðfyllandi: i.e. there is only one *stuðill* in the odd lines (as in st. 96, but always on the first lift), in contrast to stt. 80–1. All lines are type A.

85/11. The *runhent* metres are arranged in groups of three (in each group there is one full *runhenda*, one *minni*, one *minzt*), though the three in each group do not seem to have clear characteristics in common.

86/9. rétt: having the same rhyme throughout the stanza, cf. 79/13, 83/9 and n.

86/9. It is not altogether clear why this metre is said to be based on *tøglag* more than the other short-lined *runhent* metres. There are four syllables in ll. 2, 4, 5, and 8, and five in ll. 1, 3, 6, and 7. In ll. 1 and 3 the extra syllable is an enclitic (cf. 68/11–12), in l. 7 it is the result of resolution of the first lift. With *ef skjótar eru* (86/10), cf. 83/10 and st. 83 n., 70/10–12 and n. Stanza 86 is also characterized by having the two *stuðlar* in adjacent syllables near the ends of the first three odd lines (as in st. 84 and in three cases in st. 83); in st. 86 these lines are type C. It differs from st. 84 mainly in having feminine rhymes, but seems essentially the same as st. 80.

Stanza 87: cf. *Háttalykill*, st. 24 and Egill's *Hǫfuðlausn*; see JH–AH 130–1.

87/9. hneptr: but the lines still have four or five syllables, though the rhymes are masculine ones. The *stuðlar* are in syllables separated by one syllable; the odd lines are all type B (as are also ll. 2 and 8). The five-syllable lines both contain enclitics. Line 7 has six syllables; perhaps read *segk*?

88/9–10. But ll. 4 and 6 have unstressed syllables before the *hǫfuðstafr*. Lines 1, 5, and 8 have seven syllables each (*leyfi*, one of them in each case being an enclitic (*bragarmál*); cf. st. 8).

89/9. hneptr: i.e. the unstressed syllable at the end of the line is omitted. Lines 1 and 2 still have six syllables owing to resolution of the second syllables (they are presumably type A2a with secondary stress in the first dip), and l. 5 has an enclitic *ek* (read *telk*?).

Stanza 90. Lines 1, 3, and 7 have the *stuðlar* in the same position as st. 62. In l. 5 they are as in stt. 63–4. The rhythm is trochaic except perhaps in l. 1 (the first four syllables = type D). There is an extra syllable in l. 7 (enclitic *ek*) and in l. 8 (for *fyrir* read *fyr*?). The rhyme-syllables are short (thus making the end of the lines unlike stt. 62–4, but metrically equivalent to those of st. 91). The metre of this stanza is used in

Háttalykill, st. 17 (where it is called *rekit*) and in *Málsháttakvæði* (*c.* 1200; *Skj* B ii. 138–45; preserved in R after the end of *Háttatal*), alternating with that of st. 91, and generally, like st. 91, *in minzta runhenda*. It is similar to the ecclesiastical tetrameter found in medieval Latin, which may be its origin. Cf. JH–AH 131.

Stanza 91. In ll. 1–4 the *stuðlar* are in the same position as in st. 62, in ll. 5–8 as in stt. 63–4. The rhythm is trochaic throughout and there are no extra syllables (seven in every line). On the alliteration of l. 4 cf. st. 76 n. The metre of stt. 90 and 91 is equivalent, the former having the final lift in the form ⏑́ ×, the latter having it in the form –́. Both types are used in *Málsháttakvæði*.

Stanza 92. The rhythm seems to be like that of st. 83, sc. five-syllable lines, or six with one resolution or enclitic (read *hefk* in l. 2?), except for l. 3; but the penultimate syllables are all short (and probably only half-stressed; licence following long stressed syllables, see Sievers 1893, § 9.2) and the *stuðlar* come on the first and third or fourth syllables. All lines can be taken as D*2 or A2b (ll. 1–4). Cf. Sievers (1893), 115.

92/9–10. ok tekin af hálfhneptum hætti eða náhendum: in R placed at the end of the previous sentence, but it is st. 93 that is like *hálfhnept* (st. 77) and (rather less like) *náhent* (st. 75).

Stanza 93. The rhythm is like that of st. 77. Lines 1 and 3 have five syllables, l. 4 only four (cf. st. 75, *náhent*). The rest have six, in each case including a word which has a short first syllable which may be counted as constituting resolution with the second (ll. 6 and 8 also include enclitics). The *stuðlar* are placed as in st. 77 except in l. 1, where they fall on the last two syllables as in st. 84. All lines can be taken as type A2b or A*1b or A*2b (or A2ab: they all have a long half-stressed syllable in the last dip; the second word in each line except l. 4 is a finite verb, probably unstressed; cf. 77/9–10 n.).

93/9. This comment seems to refer to st. 93 (this and the misplaced clause at 92/9–10 may in the exemplar have been written in the margin, with the result that the scribe of R was unclear where they should be inserted). Though the majority of the lines in st. 93 have six syllables, they are in fact very like ll. 1–4 of st. 92 with the final two short syllables replaced by one long one. *Af fyrra hætti* would normally mean 'from the preceding verse-form' (though it is perhaps possible that it means 'from a preceding verse-form'); cf. 69/9, where the similar phrase clearly means 'in the preceding verse-form', and 91/9. Stanza 94 is also *stýfðr*, but from *hrynhenda* (stt. 62–4) or st. 90—like st. 91, with which it seems to be identical in rhythm except that it has a large number of short stressed syllables which possibly constitute resolution (ll. 3, 5, 6, and two in l. 2; cf. 62/9–12 n. and 64/9 n.); and the *stuðlar* are in the last two words in each odd line (cf. st. 62; only applies to the first half of st. 91). As in st. 91

and frequently in *hrynhenda*, there is a tendency for each line to have a finite verb in second position and to be end-stopped. Both stt. 91 and 94 are *in minzta runhenda*. The references to traditional heroes in st. 94 are reminiscent of *Háttalykill*.

Stanza 95 has double alliteration in every odd line. All lines are either type D* or type A with anacrusis except l. 5, which is type C with extended first dip. Cf. st. 79. *Fluttak fjǫgur kvæði*: cf. 69/4–5 n. *Fluttak* presumably means no more than 'composed' (and maybe sent) here, since if *Háttatal* was composed in Iceland after Snorri's first visit to Norway he would not actually have been able to *deliver* the third and fourth of his four poems to Skúli himself (in spite of the occasional use of vocatives in *Háttatal*, see 83/2 n.); the first and second would have been composed before he lęft Norway in 1220. The words cannot mean that *Háttatal* was not completed until Snorri's second visit to Norway (1237–9). The fifteen gifts Snorri received from Skúli are mentioned in *Íslendinga saga* (*Sturl.* i. 278), but the source is probably *Háttatal*.

Stanza 96 has single alliteration in odd lines. All lines are type A or type C except l. 7, which is type E. Since all the even lines are type C, the *hǫfuðstafr* is in no case on the first syllable of the line (cf. 97/10–11).

As far as st. 96, the third poem (st. 68 onwards) seems to be exclusively about Earl Skúli.

97/11. stikkalagi: probably an error for *Starkaðar lagi*, since the description fits st. 98, but cf. 97/13 textual n.: if st. 99 is *Starkaðar lag*, perhaps *stikkalag* is the correct name for the metre of st. 98. St. 99 is not, however, metrically more similar to verse attributed to Starkaðr than st. 98; and the poems *Sǫrlastikki* (*Skj* A ii. 242) and *Haraldsstikki* (*Skj* A i. 424, *Hkr* iii. 181) are not in the metre of st. 98 either. As in st. 96, the even lines in st. 98 are all type C, so that the *hǫfuðstafr* is inevitably delayed. Only occasional couplets in the verses attributed to Starkaðr in *Gautreks saga* (*Skj* B ii. 343–50) conform to the pattern of st. 98 and Snorri's description of *stikkalag*; but they are unlikely to be genuine anyway. The closest is st. 23; cf. also *TGT* 68.

97/12. sem í dróttkvæðu: i.e. there is double alliteration in the odd lines, and the *hǫfuðstafr* is at the beginning of the even lines. In st. 97 all lines are type A or type D.

Stanza 99, except for l. 1, seems identical in metre to st. 97: all lines are type A or D, there is double alliteration in the odd lines, and the *hǫfuðstafr* comes at the beginning of the even lines (l. 1 would also fit into *Bálkarlag* if it were read *Eru þeir jǫfrar* or *Þeir eru jǫfrar*). There is, however, the difference that in st. 99 all lines are kept to four syllables by the avoidance of resolutions and enclitics, which appear in four of the lines of st. 97. Stanza 99 also has virtual end-rhyme in ll. 2–8 (cf. st. 40 and 40/9 n.), though according to *TGT* 52 such grammatical rhymes were not cultivated in Norse verse, and each of these lines ends with a superlative (cf. st. 34); but these features may not be intended as part of the prescribed pattern. If stt. 99 and 94 (or 86?), which both seem to duplicate earlier verse-forms, were omitted, the poem would have exactly 100 stanzas, like a Latin *centimetrum*; cf. 100/3.

Stanza 100. Lines 1 and 4 each have only three syllables (not uncommon in eddic *ljóðaháttr*); there is only single alliteration in l. 4; ll. 2 and 5 are type B with extended first dip; there are three lifts in ll. 3 and 6.

100/1–3. This could be taken to refer to the commentary to *Háttatal*, thus confirming that it is by the poet. A hundred was the traditional number of verse-forms to include in Latin metrical treatises (*centimetra*).

Stanza 101. Line 1 has only three syllables (as sometimes in eddic *galdralag* verse); there is only single alliteration in ll. 1 and 4 (l. 2 is presumably type C, and *sótta* in that line unstressed and not alliterating, like *þá* in l. 5, which is type B). Lines 6 and 7 each have three lifts, and perhaps l. 3 too. It is, however, possible that ll. 2 and 5 are to be read with three lifts as well, with two parallel alliterations with the preceding lines (cf. *PE Sigrdrífumál* 18/5–6, 19/1–2). *Galdralag* is used occasionally in eddic poems predominantly in *ljóðaháttr*, such as *Hávamál* and *Sigrdrífumál*. *Heimdalargaldr*, of which the only known fragment is quoted in *Gylfaginning* (ch. 27), may have been entirely in this metre. Its distinguishing feature is its having a seventh line, with its repetition or echo of the sixth; cf. 8/31–2 n. Stanzas 100 and 101 are the only ones in *Háttatal* not to have eight lines (*Háttalykill*, stt. 5 and 22, like *Krákumál*, use a ten-line stanza).

Stanza 102 is *kviðuháttr*: the name is found in *TGT* 63 and in *Háttalykill*, st. 2; cf. Wessén (1915). The metre was used by Egill Skallagrímsson in both *Arinbjarnarkviða* and *Sonatorrek* (*ÍF* ii. 258–67, 246–56), by Þjóðófr of Hvinir in his *Ynglingatal* (preserved in Snorri's *Ynglinga saga*, *Hkr* i. 26–83), and by Eyvindr skáldaspillir in his *Háleygjatal* (*Skj* B i. 60–2; quoted widely in *Heimskringla* and *Skáldskaparmál*). In Snorri's example there are three syllables in odd lines (four if there is resolution) and four in even lines (or five if there is an enclitic); the first half of the stanza has single alliteration in the odd lines, the second half has double. All his odd lines have stressed initial and final syllables, whereas the earlier poets use a greater variety of rhythms. Sievers (1893, 117) suggested that *kviðuháttr* originated in the systematization of the occasional occurrence of three-syllable lines in *fornyrðislag*. The relationship, however, might be the reverse. Cf. also st. 82.

Appendix: Examples of Snorri's Metres in Other Old Norse Verse

Of the 102 variants that Snorri exemplifies in *Háttatal*, a large number are not in the strict sense of the term metrical variants: stt. 2–6 illustrate the use of kennings and are in the same metre as st. 1, which exemplifies the 'standard' *dróttkvætt* form, by far the commonest of those used by skaldic poets (*Háttalykill*, st. 3), though as Kuhn (1983), 327 points out, Snorri uses rather a restricted range of possible rhythms in his *dróttkvætt* verses, probably unintentionally, in keeping with his general tendency towards making all the lines of individual verses more consistent in pattern than was customary in earlier poets. It does not seem to be usual, however, to use kennings in a regularized way as Snorri does in stt. 2–6, so that a single stanza confines itself to one particular pattern of kenning. Nor does it seem that earlier poets took advantage of the possibilities of 'licence' provided by contraction, elision, and resolution to produce verses that seem to have a regular pattern of more or fewer than the usual six syllables in normal *dróttkvætt* as Snorri does in stt. 7–8 and in *veggjat* (st. 33), though resolution, contraction, and elision are all common enough in individual lines of verse, see Kuhn (1977); (1983), 55–6, 67–72; see also Sievers (1878). St. 7 may be inspired particularly by Eilífr Guðrúnarson's *Þórsdrápa* 5/8 (*Skj* A i. 149, *Skáldskaparmál*, ch. 18). St. 34 (*flagðaháttr*) also contains an extra syllable because of resolution that comes in superlative adjectives. This is also exemplified in *Háttalykill*, st. 32, and is found in one line of a stanza in *Víglundar saga* (*Skj* A ii. 457, st. 10; *ÍF* xiv. 104, st. 9).

Sextánmælt and *áttmælt* (stt. 9–10) are also exemplified in *Háttalykill*, stt. 21 and 38, but as patterns regularized over a whole stanza seem to be adopted from Latin poetry (JH–AH 126–7 quote as possible models verses by Matthew of Vendôme, Marbod, and other writers of the eleventh to thirteenth centuries). Few examples of these patterns being used throughout a stanza have been found elsewhere in Norse verse, though *FoGT* 151 has two stanzas in *sextánmælt* in the form of questions and answers, cf. *greppaminni* below. But although Snorri puts *sextánmælt* along with *áttmælt* and *fjórðungalok* as a rhetorical device, it in fact involves a particular type of line-rhythm which though exeptional is found quite widely in skaldic verse (in individual lines of stanzas); see Kuhn (1983), 120, 194, 268, 307, 334, who gives examples of this line-structure in individual lines of Arnórr jarlaskáld and other poets (e.g. Arnórr's *Þorfinnsdrápa* 7/5–7, *ÍF* xxxiv. 47; there are a number of examples in Hallar-Steinn's

Rekstefja, e.g. 5/3. 17/1, 6–7, 18/7, *Skj* B i. 526–9; cf. also Kormakr's *lausavísa* 37/1, *ÍF* viii. 269). Bjarni Kálfsson, *lausavísa*, *Skj* A i. 536–7 is quoted by Möbius (1879–81), ii. 129 as an example of *áttmælt*; there seems to be another in *FoGT* 134, st. 3; cf. also *Hkr* i. 367–8, stt. 165–8 and 166/5–8; *Hkr* i. 202, st. 90/3–8; Óttarr svarti, *Óláfsdrápa sœnska* 6, *SnE* (1931), 167/8–9; Hallr Snorrason 1, *SnE* 169/6–7. End-stopped lines and couplets as in stt. 10–11 are of course not uncommon in skaldic verse (they become commoner in the later period), but even *fjórðungalok* (st. 11) is not often used in a regular way to form a consistent verse-form in the way that Snorri uses it (it is found for example in Egill's *lausavísur* 3 and 27, *ÍF* ii. 109, 200; also Einarr Skúlason, *Elfarvísur* 1, *Hkr* iii. 358–9; Hallfreðr, *Óláfsdrápa* (*erfidrápa*) 1, 22/5–8, 27/5–8; *Þórðar saga hreðu*, stt. 1 (almost) and 6, *ÍF* xiv. 168, 198; *FoGT* 134, 150; see Kuhn (1969), 65–72). *Greppaminni* (st. 40; itself a form of *áttmælt*) corresponds to *Háttalykill*, st. 23, though here the full complement of internal rhymes is not included in the way that Snorri managed to do it; the only other parallels to this in Norse verse may all be later than *Háttalykill* (see JH–AH 125–6; Vésteinn Ólason 1969; 40/1–4 n.). Similar rhetorical devices are found in Latin verse.

Parenthetical sentences such as Snorri uses in *stælt* (st. 12) are common in skaldic verse, but it is not all that common for them to extend over the whole of ll. 2–3 of a half-stanza; cf. Hallfreðr's *Óláfsdrápa* (*erfidrápa*) 6/5–8, 20/5–8, 25/1–4, *Skj* A i. 160, 163, 165, B i. 151, 154, 156; Einarr Skúlason, *Geisli* 1/1–4, *Skj* B i. 427; *FoGT* 136–7 can only exemplify this from *Háttatal*, but there are a surprising number of examples in Þorkell Gíslason's *Búadrápa*: stt. 1, 3/5–8, 4, 7/5–8, 8, 12/1–4, *Skj* B i. 536–8. For *hjástælt* (st. 13), however, there was a well-known model in Kormakr's *Sigurðardrápa* (*Skj* A i. 79–80, B i. 69–70; preserved in quotations in *Skáldskaparmál* and *Heimskringla*); this seems to be the only example of the use of *forn minni* (though some lines in the Norwegian Runic poem, *Skj*. B ii. 248–9, are comparable), which according to Snorri is essential to the form, but examples of the same line-structure are not uncommon, and for instance Snorri's stt. 26–7 (*orðskviðuháttr* and *álagsháttr*) and *Háttalykill*, st. 40 also use it; cf. Haraldr harðráði's *lausavisur* 15–16, *Skj* A i. 359–60, B i. 331 and various further examples of individual lines with this pattern in Kuhn (1983), 179–82 (cf. also 289; proverbs are not normally used, however, and examples with the pattern in every couplet of a stanza as in *Háttatal*, stt. 26–7 and *Háttalykill*, st. 40 are not found). See also JH–AH 131–2 and Kuhn (1969*c*), 72, who compares Þjóðólfr, *Haustlǫng* 17/1–4, *Skj* B i. 17. The structural pattern of *tilsagt* (st. 25; *Háttalykill*, st. 34, '*tilsegjandi*') is also quite common (Kuhn 1983, 167–8, 194–5); many, though not all, parenthetical sentences in this position in the verse are explanatory as in Snorri's example (in Egill's *Hǫfuðlausn* 8/4, *ÍF* ii. 188, the metre is not *dróttkvætt*; but cf. Egill's *lausavisa* 42/4, *ÍF* ii. 274, and *Skj* B i. 399, st. 6/7). *Langlokum* (st. 14) seems to be exemplified only in *Háttalykill*, st. 30, though the device is similar to that of the *klofastef*, stt. 68–70; cf. *FoGT* 136, Fidjestøl (1982), 249, JH–AH 133–4. On the name see 14/1 n.; it is presumably derived from a phrase such as *yrkja langlokum*.

Syntactical linking of different stanzas, especially grammatical subordination of parts of a verse to a preceding one (*tiltekit*, st. 15) is generally avoided by skaldic poets (Kuhn 1969*c*, 63), but because so few connected stanzas from skaldic poems have survived, it is difficult to tell how often it occurs; if the first extant line of a stanza seems to be dependent on a preceding line this is usually taken to mean that a preceding line from the same stanza has been lost. The only clear example seems to be Einarr Skúlason's *Geisli* 1–2 (*Skj* A i. 459, B i. 427; there are examples also in *Sonatorrek*, see *ÍF* ii. 247, note on st. 3, but this is not *dróttkvætt*). Stanzas beginning with *ok*, *unz*, or *áðr* do, however, occur (Bragi, *Ragnarsdrápa* 8, 11, *Skj* A i. 2–3; Þjóðólfr, *Haustlǫng* 11, 20, *Skj* A i. 18, 20, B i. 16, 18; Einarr skálaglamm, *Vellekla* 7, 16, *Hkr* i. 208, 242; cf. also *Háttatal*, stt. 39 and 88). Syntactical linking of the two halves of a stanza (st. 39) is less uncommon (examples from Þorbjǫrn hornklofi's *Glymdrápa* are given in Fidjestøl (1982), 218; see also Kuhn (1969*c*), 63–4, 72, who quotes Hallfreðr, *Óláfsdrápa* (*erfidrápa*) 29. *Drǫgur* (st. 16) seems to be commoner outside Norse poetry (in Old French and Middle English, see 16/9–10 n.); *TGT* 94 is only able to offer Snorri's example in illustration of this kind of verse-linking (which is there called *anadiplosis*), but of course more examples might be apparent in skaldic verse if we had more complete series of stanzas preserved. *Refhvǫrf* (stt. 17–23) appear also in *Háttalykill*, stt. 20, 28, and 35, where the antitheses are called *refrún*; this device is also found widely in medieval Latin verse (JH–AH 119, 124–5). Snorri also uses it as an occasional embellishment elsewhere (*Háttatal*, st. 44/8), but its frequency in Norse verse hardly justifies the amount of space (with his careful subdivisions of types where it is used in regular patterns within stanzas) that he devotes to it in *Háttatal*. The type represented by Snorri's *síks glóðar* (17/1) of course appears in all kennings for gold of the type 'fire of the sea'; the type involving *ofljóst*, such as *heit kǫld* (17/4) is hard to parallel (cf. Meissner 1921, 83; de Vries (1964–7), ii. 31).

A second type of verse-form that Snorri has a number of examples of is also not strictly metrical, but involves special arrangements of the hendings. *Dunhenda* (st. 24; *Háttalykill*, st. 33), *samhent* (st. 46), *iðurmælt* (st. 47; *Háttalykill*, st. 29, though with less regular hendings) and *klifat* (st. 48; cf. *Háttalykill*, st. 12) involve repetition of complete syllables either within the line or from one line to another; *liðhendur* (stt. 41 and 53) involves assonance where syllables both alliterate and have *skothending*, and also continues the rhyme in the even lines. There are examples of these devices in skaldic verse, though the particular types of repetition Snorri enumerates are rarely if ever maintained with complete regularity throughout a stanza. The development of their use is traced by Hans Kuhn (1981); they are particularly common in the verse of Egill (see his *lausavísur* 4, 16, 22/3, 41/1, 47, *ÍF* ii. 110, 156, 170, 269, *TGT* 86), Þorbjǫrn hornklofi (*Glymdrápa* 4/1–2, 6, 8/1–4, *Hkr* i. 103, 112–13, 121; hence perhaps the poem's name; cf. Fidjestøl 1982, 219–21), Þjóðólfr of Hvinir (*Haustlǫng* 9/1–2, 12/7–8, 19/3–4, *Skáldskaparmál*, chs. 17, 22) and Einarr skálaglamm (*Vellekla* 1, 5/3–4, 15/1–2, 36, *Skj* A i. 122–3, 125, 131). Their frequency declines in later poets, who more and more adopted the rigid pattern of regular *dróttkvætt*, but examples occur in Eilífr Guðrúnarson, *Þórsdrápa* 16, *Skj* A i. 151; Kormakr, *lausavísur* 5/7–8, 10/7–8, 17/1–2, 5–6, 27/1–2, 28/1–4, 31/7–8, 35/5–6, 38/7–8, 44/7–8, 47/5–6, 60/3–4, *ÍF* viii. 211, 214–15, 221, 242–3, 245, 268, 271, 276, 283, 289; Úlfr Uggason, *Húsdrápa* 3/1–2, *Skj* A i. 137; Glúmr Geirason, poem on Eiríkr blóðøx 1, *TGT* 94; his *Gráfeldardrápa* 3/3–4, *Hkr* i. 162; Halldórr ókristni, *Eiríksflokkr* 1/3–4, *Hkr* i. 350; Óttarr svarti, *Hǫfuðlausn* 2, 4/3–4, 10/7–8, *Hkr* ii. 6, 20; Eyvindr skáldaspillir's *lausavísa* 9/1–2, *Skj* A i. 73; Sighvatr Þórðarson, *Austrfararvisur* 14/1–2, *Hkr* ii. 139; Þórðr Kolbeinsson, *Eiríksdrápa* 11/1, *Skj* A i. 216 (only in AM 61 fol.); Gísli Súrsson's *lausavísa* 30, *ÍF* vi. 105–6; Hallfreðr's *lausavísa* 11, *Hkr* i. 331, *ÍF* viii. 161–2, and his *Óláfsdrápa* 9, *Hkr* i. 265; Bjǫrn Hítdœlakappi's *lausavísa* 19/1–2, *ÍF* iii. 171; Tindr Hallkelsson's *drápa* on Earl Hákon 1/1–2, *Hkr* i. 281 and his *lausavísa* 1/1, *ÍF* iii. 308; Guthormr sindri, *Hákonardrápa* 6/5, 7–8, *Hkr* i. 174; Haraldr harðráði's *lausavísa* 19/3–4, *Hkr* iii. 188; his *lausavísa* 10 together with Þjóðólfr Arnórsson's *lausavísa* 12/5–6, *Skj* A i. 358, 379; Þjóðólfr Arnórsson, *Magnúsflokkr* 17/5, *Hkr* iii. 53; *Gunnlaugs saga*, stt. 23–5, *ÍF* iii. 104–7; Einarr Skúlason, *Skj* A i. 480–1, stt. 6 and 8, *SnE* (1931), 172/10–11, 173/3–4; *Skj* A ii. 209–10, B ii. 225, st. 7/5–8; *Skj* A ii. 213, B ii. 230, st. 2/3, 7; *FoGT* 134, stt. 1–3. Note also Snorri's own *lausavísa* 7, *SnE* (1924), 112. In some early skaldic verse such devices between lines seem sometimes to replace regular hendings within lines (e.g. Egill's *Skjaldardrápa*, *ÍF* ii. 272–3; Bragi, *Ragnarsdrápa* 4/7–8, 5/5–6, 9/7–8, 11/5–6, 13/3–4, 16/3–4, *Skj* A i. 2–4; cf. Kuhn (1981); (1983), 87–8, 292; see also *Háttatal*, st. 58, *Braga háttr*, below). There are also many similar devices used in medieval Latin verse, see JH–AH 127–9, though it is clear that for Snorri himself there were plenty of precedents in skaldic verse (particularly that by early poets whom he seems to have particularly liked, such as Bragi, Egill, and Einarr skálaglamm), whatever their ultimate origins. These figures regained popularity in the later Middle Ages, perhaps under the influence of Snorri's treatise: there is a late example of *dunhenda* from a 'tjald í Hólakirkju' printed in *SnE* (1848), 248, and *dunhenda* is combined with *runhenda* in a verse in *Harðar saga*, ch. 38, *Skj* A ii. 449; the resulting pattern is similar to that of the couplet of Glúmr

Geirason, *TGT* 94, and Hjalti Skeggjason's *kviðlingr*, *Skj* A i. 139. Cf. Meissner (1921), 83 n.

The particular way of positioning the alliterating staves in *tvískelft* (st. 28) and *in forna skjálfhenda* (st. 35, cf. st. 39/10–11) with *skjálfhendur* is quite common in skaldic verse in individual lines, but it is used with regularity only in Hallvarðr háreksblesi, *Knútsdrápa*, *Skj* B i. 293–4, which to judge from Snorri's account of the form (see 35/13 n.) is predominantly in *in nýja skjálfhenda* (see Fidjestøl 1982, 125), and Hallar-Steinn's *Rekstefja*, *Skj* B i. 525–34, which is the only consistent example of *tvískelft* (the name of the form and the rarity of it are both mentioned in st. 35 of the poem); cf. Kuhn (1983), 74, 104–6, 268, 289 (examples in odd lines of Kormakr's verse), 333–4. Cf. *Háttalykill*, st. 41 (to judge from Snorri's commentary this also is *in nýja skjálfhenda*, cf. notes to 28/14 and 35/13). There do not seem to be examples of poems consistently using *in forna skjálfhenda*.

The rhythmical pattern of *detthendr* (st. 29) is also found in *Háttalykill*, st. 18 (but with a different placing of rhymes); but in skaldic verse generally, while it is not uncommon for a b-line to begin with a trisyllabic word, it is not very usual for another trisyllable to follow—in the vast majority of *dróttkvætt* verses all lines end with a disyllable. A few examples are given in Kuhn (1983), 177–8 (among which are Þjóðólfr of Hvinir, *Haustlǫng* 1/3, 13/8, 20/8, *Skj* B i. 14, 17–18, *Skáldskaparmál*, chs. 17 and 22; and Hallfreðr, *Óláfsdrápa* (*erfidrápa*) 29/2, *Skj* B i. 156). The pattern seems to occur particularly frequently in Sighvatr's verse: *Víkingarvísur* 8/8, 9/8, *Hkr* ii. 20–1; *Nesjarvísur* 5/4, *Hkr* ii. 61; *flokkr* on Erlingr Skjálgsson 9/8, *Hkr* ii. 28; poem on Queen Ástríðr 3/4, *Hkr* iii. 6; *Bersǫglisvísur* 8/4, *Hkr* iii. 28. See Sievers (1893), 99–100: most of these examples are to accommodate trisyllabic names into the verse. *Draugsháttr* (st. 30, *Háttalykill*, st. 36) simply regularizes one of the commonest patterns of even lines in skaldic verse (type A, leading to a regular trochaic rhythm; Kuhn (1983), 162), though it is not easy to find examples of verses which have this pattern in all four even lines. *Bragarbót* (st. 31) appears rarely in skaldic verse (e.g. Kormakr, *Sigurðardrápa* 7/3, *Skáldskaparmál*, ch. 2; see Kuhn (1983), 147; (1969*a*), 228–30) and certainly never consistently throughout a stanza. *Riðhendur* (st. 32, cf. also stt. 55–6) also regularizes a rare type of even line which according to Kuhn is not used by the strict court poets, though there are examples in Bragi, Egill, and Kormakr (1983, 165; 1969*a*, 228–9; another example in *TGT* 98). *Stamhendr háttr* (st. 45) seems to have its closest parallel in Kormakr's *lausavísur* 5/7, 30/7, 38/1, 39/1, and 49/7, *ÍF* viii. 211, 245, 270–1, 285 (Kuhn 1969*a*, 226–7; 1983, 157, 289); cf. also Hallfreðr, *Óláfsdrápa* (*erfidrápa*) 8/1, *Skj* B i. 152; Haraldr harðráði, *lausavísa* 13/3, *Hkr* iii. 134 (it is rare to find examples with *aðalhending*).

Of the other variations to the internal rhyme-scheme, *þríhent* (st. 36) is also exemplified in *Háttalykill*, st. 6, *hinn dýri háttr* (st. 37), and the unnamed verse-form of st. 38 in *Háttalykill*, st. 9 (called *hinn dýri háttr* but actually more similar to *Háttatal*, st. 38), but these forms do not seem to have been used by skaldic poets. A single possible example of a line with three *aðalhendingar* is found in Bragi's *Ragnarsdrápa* 13/2, *Skj* A i. 3, depending on the reading adopted; cf. 4/1 and 5/3, which each have three *skothendingar*; note also Kormakr's *lausavísur* 39/5 and 60/3 (*ÍF* viii. 271, 298) and Hallar-Steinn's *Rekstefja* 18/2, *Skj* A i. 547. There are similar things in Latin verse (JH–AH 129–30). Examples of *aðalhending* in odd lines as well as even lines of verses as in *rétthent* (st. 42) are found in skaldic verse, but not often maintained throughout a stanza: Bragi, *Ragnarsdrápa* 4/3, 11/7, *Skj* B i. 1, 3; Egill's *lausavísa* 10/1–4, *ÍF* ii. 142; Kormakr, *lausavísur* 15/3–4, 33/5–6, 37/3–4, 7–8, 50/5–6, 54/5–6, 55/7–8, 59/7–8, 60/1–2, 63/7–8, *ÍF* viii. 219, 265, 270, 286, 290–1, 297–8, 301; Einarr skálaglamm's *Vellekla* has many examples (see Kuhn 1981, 304); Bjǫrn krepphendi, *Magnúsdrápa* 2, *Hkr* iii. 217; Magnús berfœttr, *lausavísa* 3/1–4, *Skj* B i. 402; Hallfreðr, *Óláfsdrápa* (*erfidrápa*) 2/5–3/4, *Hkr* i. 356–8; Hallbjǫrn Oddsson, *lausavísa*, ll. 1–4, *ÍF* i. 193; Þorbjǫrn Brúnason, *lausavísa* 1/1–4, *ÍF* iii. 289; *TGT* 99, st. 3/1–4; see Kuhn (1983), 87. There are a few examples of skaldic poets using two pairs of rhymes in a single line (*alhent*, *in minni alhenda*, stt. 43–4), besides the verse of Klœingr quoted by Snorri: Bragi, *Ragnarsdrápa* 8/4, 17/4, *Skj* B i. 2, 4; Hofgarða-Refr in his poem on Gizurr Gullbrárskáld, st. 3, *Skáldskaparmál*, ch. 2 (see Kuhn 1983, 305–6); Hallar-Steinn, *Rekstefja* 18/1, *Skj* B i. 529; cf. also *Sturl.* i. 278 (the *stef*, but probably not the whole poem, of one of Snorri's otherwise lost eulogies of Earl Skúli, was *alhent*); *FoGT* 124. *Skothendr háttr* (st. 52), where *skothendingar* are used in even as well as odd lines, is found frequently in odd couplets of *dróttkvætt*, particularly in earlier poets, but rarely consistently throughout a stanza, e.g. Bragi's *Ragnarsdrápa* 5/1–2, 7–8, 6/7–8, 9/1–2 (see Kuhn 1983, 87; Kormakr's *lausavísur* contain several examples). *Ragnars háttr* (st. 54), *Torf-Einars háttr* (st. 55), *Egils háttr* (st. 56) and *munnvǫrp* (st. 66, *Háttalykill*, st. 8) have no rhymes in odd lines (and in stt. 55 and 66 the even lines have only *skothending*); such lines are common in earlier skaldic verse and in verse in *fornaldar sögur*, though not generally used consistently throughout a stanza, see notes to these stanzas, and Kuhn (1981), 303–4: there are many cases in Einarr skálaglamm's *Vellekla*; Kuhn (1983), 87–9, 292, 326; there are also a number of examples in Kormakr's *lausavísur*. None of the extant verses of Ragnarr, Torf-Einarr, or Egill corresponds exactly to the metres Snorri names after them, though of course Snorri may have known other poetry by them that has not survived; Kuhn (1983), 326 points out that Snorri gives stt. 55–8 two characteristics each, but that these are not often in fact found together in extant verse. Cf. Finnur Jónsson (1920–4), ii. 84–5; *Njála* 1875–89, ii. 17–22. On *riðhendur* in stt. 55–6 see above; unstressed syllables at the beginning of even lines as in *Ragnars háttr* occur four times in Bragi's *Ragnarsdrápa*, see Kuhn (1983), 168–9. *Munnvǫrp* is, however, used fairly consistently in Bjarni Kolbeinsson's *Jómsvíkingadrápa*, *Skj* B ii. 1–10, the text of which is included at the end of *Háttatal* in R (Finnur Jónsson 1920–4, ii. 41); also in Kormakr, *lausavísur* 9/7–8, 13/1–4, *ÍF* viii. 214, 217; Bárðr á Upplǫndum, *lausavísa*, *Skj* B i. 145.

Fleins háttr (st. 57) has the second hendings on the second lift instead of the last in each line; so has st. 58. Occasional lines of this type occur in the very earliest skaldic verse: Bragi, *Ragnarsdrápa* 5/7, 8/6–7, 10/7, 16/1, 20/2, *Skj* B i. 2–4; two lines in the verse in *Egils saga*, ch. 71, *ÍF* ii. 224–5 (see Kuhn 1981, 306–7). But in general the rule that the second hending must fall on the penultimate syllable of the line is strictly observed (cf. 1/38–9; Kuhn 1983, 89–90, 135, 162). The essential feature of *Braga háttr* (st. 58) is that the odd lines lack internal rhyme, but one syllable in them rhymes with one or more syllables in the even lines. This feature is quite common in early skaldic verse (though never used consistently throughout a stanza), including Bragi's (see note to st. 58), Egill's, and Einarr skálaglamm's, e.g. Egill's *Skjaldardrápa* and *lausavísur* 27/1–4, 30/1–2, 42/1–4, 43/7–8, *ÍF* ii. 200, 204, 272–4, 293; Einarr skálaglamm's *Vellekla* 5/3–4, 10/3–4, 12/3–4, 28/1–2, *Skj* A i. 123–5, 129; also Þjóðólfr of Hvinir's *Haustlǫng* 7/5–6, 14/1–2, *Skj* A i. 17, 19; Kormakr's *lausavísur* 6/3–4, 13/5–6, 21/7–8, *ÍF* viii. 211, 217, 229; see Kuhn 1981; 1983, 87–8, 292. There are, however, few if any cases where there are *skothendingar* in the even line, one of which forms *aðalhending* with the rhyming syllable in the odd line (the closest are perhaps Þjóðólfr of Hvinir's *Haustlǫng* 13/7–8, 20/7–8, *Skj* A i. 19–20, though these two couplets each contain four rhymes; and Egill's *lausavísa* 30/1–2, depending on the readings adopted), and none where the hendings in the even line are on the first two lifts; but in the sole surviving verse of Egill's *Skjaldardrápa* (*ÍF* ii. 272–3) there are hendings between odd and even lines in three out of the four couplets, and other varieties of the pattern may have occurred in other verses of the poem. In another verse in *Egils saga* (*ÍF* ii. 224–5), which although not attributed to Egill may in fact be by him, there are rhymes between odd and even lines and also lines with the two hendings on the first two lifts, but not in the same couplet (unless this occurs in ll. 3–4); cf. Bragi's *Ragnarsdrápa*, stt. 5 and 16 (*Skj* A i. 2, 4). These kinds of rhymes are rare after the end of the tenth century (Kuhn 1981, 306).

Háttlausa (st. 67; *Háttalykill* st. 26, with more regular placing of *hǫfuðstafr*), where hendings are entirely lacking, is found occasionally in early verse too (there are a number of examples in Egill's *lausavísur*, e.g. most of nos. 1, 3, 7, 38, 40, *ÍF* ii. 100–1, 109, 121, 230, 268; also Kormakr, *lausavísa* 13/7–8, *ÍF* viii. 217; *Hallfreðar saga*, *ÍF* viii. 142, *Skj* B i. 168–9; many of the verses in *Ragnars saga* and other *fornaldar sögur*). *Krákumál*, which only has sporadic hendings, uses a ten-line stanza that Snorri does not include in *Háttatal* (cf. Finnur Jónsson 1920–4, ii. 152–3). Egill's *lausavísur* contain a large number of variations from the standard hending pattern of *dróttkvætt*, though he rarely uses them consistently throughout a verse, and many of these, together with those of Bragi's *Ragnarsdrápa*, the metre of which also varies considerably from the later norm, may have been the source of Snorri's variant forms that regularize the use of such variations; the work of both poets was clearly well known to Snorri. Kormakr's *lausavísur* also show a great deal of freedom in the arrangement of hendings, and another poet that uses many such variations is Einarr skálaglamm (Kuhn 1981; 1983, 291–3). It is from the regularization and systematization of such sporadic variations in earlier poets, in whose practice the strict rules that became characteristic of later *dróttkvætt* did not yet apply, that many of Snorri's forms are created.

There are actually comparatively few of Snorri's variants of *dróttkvætt* that involve lines with more or fewer than six syllables where these are not accounted for by *leyfi* (stt. 7–8, 33–4). His three *stúfar* (stt. 49–51; cf. *Háttalykill*, stt. 5, which seems to have ten lines in the stanza, cf. JH–AH 44, and 31) have no real precedents in Norse verse, though the pattern of *inn mesti stúfr* is actually almost identical to that of *hálfhnept* (st. 77, see below) and many of the poems taken here to be examples of the latter could equally well be considered to be *stýft*; cf. 77/12 and see Kock (1933). To judge from Snorri's descriptions, the only real difference is that the line-length in the *stúfar* was fixed, while it was variable in *hálfhnept*, and the rhythms in Snorri's *stúfar* are more like *dróttkvætt* with the final syllable omitted (mostly types D and E), while in st. 77 the rhythms of the lines as wholes are more like Sievers types A2 and A*2. There does however seem to be an isolated example of *inn meiri stúfr* (st. 50) in Þórir snepill's *lausavísa*, *ÍF* i. 270. It has been suggested that *stýft* developed under the influence of Latin verse (see JH–AH 120). There is also a possibility that the idea of catalectic lines originated in corrupt readings, e.g. Þorleifr jarlsskáld, *lausavísa* 2/8, *ÍF* ix. 174, and the verse of Brynjólfr úlfaldi, *Hkr* ii. 82. *Kimblabǫnd* (stt. 59–61) are in Norse verse before Snorri apparently only found in *Háttalykill*, st. 14, but examples are found in medieval Latin and French verse (JH–AH 129), and the influence of Irish verse has also been proposed (Heusler 1925, 313). Rhythmically, of course, *in mestu kimblabǫnd* (st. 61) is like *hrynhenda* (stt. 62–4), *dróttkvætt* with a trochee added at the end of each line.

Hrynjandi or *hrynhenda* (stt. 62–4) became one of the most fruitful of the developments of *dróttkvætt* that almost certainly had its origin in medieval Latin verse; on this metre see Kuhn (1983), 312, 337–41. Snorri's three variants are not used consistently in earlier examples, but the eight-syllable line appears, seemingly for the first time, in *Hafgerðingadrápa* (*Skj* B

i. 167, *ÍF* i. 132–4, 395), *c*.986, apparently from the Hebrides, and Christian in content (though the traditional early dating has been questioned, see Jakob Benediktsson 1981); its best-known early exponent is Arnórr jarlaskáld in his *Hrynhenda*, *Skj* A i. 332–8 (composed in 1046). It was also used by Markús Skeggjason in *Eiríksdrápa*, *Skj* A i. 444–52, Gamli kanóki in *Jóansdrápa*, *Skj* A i. 561, in two anonymous verses in *Sverris saga* (*Skj* B i. 596–7), in a verse in one version of *Fóstbrœðra saga* (*ÍF* vi. 233–4 n. 5), by Óláfr hvítaskáld, Sturla Þórðarson, and many later poets (see *Skj* B ii. 610). In the fourteenth century it became a favourite medium for religious poems; the best-known is Eysteinn Ásgrímsson's *Lilja* (*Skj* B ii. 390–416). It is represented in *Háttalykill* by st. 16 (cf. JH–AH 120–1), while the variant in st. 63 (*trollsháttr*) has a near equivalent in *Háttalykill*, st. 37 (*konungslag*). *Háttatal*, st. 90 is an end-rhyming version of the form (though with a short final lift) which shows even closer affinity to the Latin hymnic trochaic tetrameter and is also found in *Háttalykill*, st. 17, where it is called *rekit*; this is also in part the metre of *Málsháttakvæði*, *Skj* A ii. 130–6, included at the end of *Háttatal* in R (see JH–AH 131; Murphy 1961, 21 ff.). It is noticeable how all versions of *hrynhenda* tend to end-stopping and a trochaic rhythm based solely on stress, with lifts falling more and more on short syllables (see 93/9 n.). The seven-syllable line of *draughent* (st. 65) corresponds to *Háttalykill*, st. 4, but does not seem to be found elsewhere.

Fornyrðislag, with its variants *Starkaðar lag* (or *stikkalag*; this metre does not correspond closely to that of verses attributed to Starkaðr, see 97/11 n.) and *Bálkarlag* (stt. 96–9, *Háttalykill*, st. 19) is in fact the commonest metre used in the poems in *Skj* i–ii (see *Skj* B ii. 609) after *dróttkvætt*, though actually not many of the poems in such metres are court poems in praise of kings and earls, the main ones that are being the poem attributed to Þjóðólfr in *Flateyjarbók* on Haraldr hárfagri, *Skj* B i. 18–19; the fragment of a poem probably about Haraldr harðráði by Sneglu-Halli, *Skj* B i. 358; *Haraldsstikki*, *Skj* B i. 394; Gísl Illugason's poem on Magnús berfœttr, *Skj* B i. 409–13; Ívarr Ingimundarson's *Sígurðarbálkr*, *Skj* B i. 467–75; cf. Fidjestøl (1982), 177; Kuhn (1983), 312, 337. Besides its use in eddic and eddic-type verse (e.g. *Darraðarljóð* and Gunnlaugr Leifsson's *Merlínússpá*, *Skj* B i. 389–91, ii. 10–45) and riddles, it is found frequently in *fornaldar sögur*, where it is often attributed to legendary and supernatural characters, though it can be used by skaldic poets (e.g. Egill's *lausavisa* 46, *ÍF* ii. 296) in *lausavísur*, particularly those of a less formal kind, as well as in *kviðlingar* and *níð*; and a number of *fornyrðislag* verses are attributed to kings, such as Haraldr harðráði and Óláfr helgi. (The verse in *TGT* 68 is there said to be in *Bálkarlag*, but does not follow the pattern of *Háttatal*, st. 97 exactly.) The other so-called eddic metres, *málaháttr* (st. 95; used sporadically in eddic verse alongside *fornyrðislag*, but consistently in

Atlamál) and *ljóðaháttr* (st. 100, *Háttalykill*, st. 1) are less common, but are used in some important poems, often alternating with each other or with *fornyrðislag* within the same poem: Þorbjǫrn hornklofi's *Haraldskvæði* or *Hrafnsmál*, *Skj* B i. 22–5; Eyvindr skáldaspillir's *Hákonarmál*, *Skj* B i. 57–60; *Eiríksmál*, *Skj* B i. 164–6; *Bjarkamál*, *Skj* B i. 170–1; as well as *Hugsvinnsmál*, *Skj* B ii. 185–210 and some verses in *fornaldar sögur*. Cf. Fidjestøl (1982), 176; Heusler (1925), 230–2; Kuhn (1983), 336; on the names see Wessén (1915), 129–34. *Galdralag* (st. 101), however, seems only to be found in eddic poems predominantly in *ljóðaháttr*, such as *Hávamál* and *Sigrdrífumál* (*Skírnismál*, *Grímnismál*), as an occasional variation (see Heusler 1925, 247–9), though an extra line similar to the seventh line of a *galdralag* verse is often found in *dróttkvætt* dream-verses, see 8/31–2 n.

Of the other forms with short lines that may be regarded as variations of *fornyrðislag*, one of the best-known is the rhymeless *kviðuháttr* (st. 102, *Háttalykill*, st. 2, where the name, lacking in *Háttatal*, is given; see also *TGT* 63), distinguished by its three-syllable odd lines (cf. Kuhn 1983, 336). It is used in Þjóðólfr of Hvinir's *Ynglingatal*, *Hkr* i. 26–83; Egill's *Sonatorrek*, *Arinbjarnarkviða*, and *lausavísa* 25, *ÍF* ii. 246–56, 258–67, 193–4; Eyvindr skáldaspillir's *Háleygjatal*, *Skj* B i. 60–2; Grettir's *Ævikviða*, *Hallmundarkviða*, and other verses in *Grettis saga*, *ÍF* vii. 86–7, 170–2, 176–7, 203–4; Þórarinn loftunga's *Glælognskviða*, *Skj* B i. 300–1, *Hkr* ii. 399, 406–8; *Noregskonungatal*, *Skj* B i. 575–90; Sturla's *Hákonarkviða*, *Skj* B ii. 118–26; verses in *Stjǫrnu-Odda draumr*, *Skj* B ii. 222–3; and elsewhere; cf. also Fidjestøl (1982), 175–7. The element *-kviða* (when it refers to skaldic rather than eddic poems) is usually reserved for titles of poems in this metre, though Halldórr skvaldri's *Útfararkviða* (*Hkr* iii. 245) is in fact another *fornyrðislag* poem (the name seems to be a modern one in any case). In view of the content of the poems in this metre, there seems little reason to connect either the name *kviðuháttr* or the element *-kviða* with the word *kvíða* '(express) anxiety (about something), lament'; cf. Hallvard Lie in *KLNM* ix. 559 ('Kviðuháttr') and Wessén (1915).

The other short-line forms all involve either internal or end-rhymes. *Tøglag* or *tøgdrápulag* (stt. 68–70, *Háttalykill*, st. 13) may be named after Þórarinn loftunga's *Tøgdrápa*, *Skj* A i. 322–4, *Hkr* ii. 308–10, which uses this form, as do Sighvatr's *Knútsdrápa*, *Skj* A i. 248–51—both these poems had a *klofastef* like Snorri's example, and both are concerned with journeys—Einarr Skúlason's *Haraldsdrápa* II, *Skj* A i. 457–8, and Þórarinn stuttfeldr's *Stuttfeldardrápa*, *Skj* A i. 489–91; it is also used in two fragments in *TGT* 83 and 95 (cf. *Háttatal*, st. 71), see Fidjestøl (1982), 176–7; cf. also Kuhn (1983), 300, 304, 312. Outside these poems the form is used rarely (Ámundi Árnason's *lausavísa* 5, *Sturl.* i. 314; an anonymous half-stanza in *Sturl.* i. 217), though there is an irregular example also in the pair of stanzas, one of them attributed to Bragi, in *Skáldskaparmál*, ch. 54. *Háttalykill*, st. 13, like *Tøgdrápa*, *Knútsdrápa*, and *Stuttfeldardrápa*, sometimes has hendings in odd lines (*Háttatal*, stt. 68 and 70) and sometimes has two *stuðlar* (*Háttatal*, stt. 69 and 70), but Einarr Skúlason follows the pattern of *Háttatal*, st. 70 consistently.

Hinn grœnlenzki háttr, *hinn skammi háttr*, *nýi háttr*, *stúfhent* (stt. 71–4) all have four-syllable lines with various arrangements of hendings, but do not seem to be used by skaldic poets (cf. Finnur Jónsson 1920–4, ii. 85; JH–AH 129: they may be based on foreign metres), though all four could in fact be taken as variants of *tøglag*; for instance the pattern of *stúfhent* is found in 69/6 and 70/2, 4 and *hinn grœnlenzki háttr* is very like *tøglag*, but has two-syllable rhymes in the even lines, cf. *TGT* 83 and 96. *Háttalykill*, st. 10 is also similar to *Háttatal*, st. 71 (though in it not all the odd lines have *skothending* as Snorri's have), and two verses in *FoGT* (137–9) use *hinn nýi háttr* (this form may have been invented by Snorri; his example only is quoted in *TGT* 98). *Náhent*, *hnugghent*, *hálfhnept*, *alhnept* (stt. 75–8) also involve short lines, mostly with rather different rhythms from *tøglag*, but apart from *Háttalykill* (of which st. 15 is similar to *Háttatal* 75, though less regular, and stt. 25, 35, 39 to *Háttatal*, stt. 77 and 78; cf. JH–AH 119, 125, 130), the only real precedents seem to be for *hálfhnept* (*Háttalykill*, st. 25, though here the even lines are actually more like *Háttatal*, st. 78). The most notable is Óttarr svarti's *Óláfsdrápa sœnska*, *Skj* A i. 289–90 (all the fragments of which are found in Snorri's *Edda* only; see Fidjestøl 1982, 171, 177), but the form also seems to be used in Haraldr hárfagri's *Snæfríðardrápa*, *Skj* B i. 5 (according to Ólafur Halldórsson 1969 this fragment in fact belongs with Ormr Steinþórsson's poem below, and was probably composed in the twelfth century); Bjǫrn Breiðvíkingakappi's *lausavísa* 5, *ÍF* iv. 110–11; Brynjólfr úlfaldi's *lausavísa*, *Hkr* ii. 82, cf. n. 1; Haraldr harðráði's *lausavísa* 8 and Magnús góði's *lausavísa* 1, *Skj* A i. 330, 358; Ormr Steinþórsson's poem apparently about a woman (Snæfríðr?), *Skj* A i. 415–16, *Skáldskaparmál*, chs. 3, 47, and 61, and Faulkes (1977–9), i. 397 (see Poole 1982); the two couplets in *Kormaks saga*, *ÍF* viii. 216; the verse attributed to Magnús góði in *ÍF* xi. 331; Eilífr Snorrason, *lausavísa* 1, *Skj* A ii. 42–3; Gunnarr, *lausavísa*, *Sturl.* ii. 109; *Skj* A i. 600, st. 34, Faulkes (1977–9), i. 377; and is occasionally used in later poems too, e.g. Árni Jónsson, *lausavísa* 2, *Skj* B ii. 461; note particularly *Máríuvísur* II–III, *Skj* B ii. 532–45; see Kock (1933). Snorri's st. 77 is, however, very similar to st. 51, as noted above (both have lines of 5–6 syllables, though *Háttalykill*, st. 25 mixes four, five, and six syllable lines), and it is difficult to tell which type some of the above poems are closest to; many of them have very irregular rhythms, and occasionally have the four-syllable lines of *Háttatal*, st. 78 (e.g. Óttarr's *Óláfsdrápa* 2/2, 3/2, 4/1, 3, 6/1); some have occasional lines perhaps more similar to *Háttatal*, st. 75. Most of them are listed in *Skj* B ii. 610 as *stýft*, and the editor's note in *ÍF* xi. 331 identifies the form of Magnús góði's supposed verse as *inn mesti stúfr*. Snorri himself remarks on the similarity (77/12) and on the variability in length of lines in *hálfhnept* (77/9–10). It is possible that in stt. 75–8 Snorri has systematized variations found sporadically in such poems as those listed above, making several subtypes out of the one traditional verse-form, as he has done for instance with *tøglag* and *hrynhenda*. The four verses are all characterized by the inclusion of lines ending with two heavy syllables (of which the first is sometimes resolved, as in st. 76; alternating with lines ending in a trochee in stt. 75–6). *Haðarlag* (st. 79; *Háttalykill*, st. 27 is similar but with less regular hendings) is easier to identify, and is found in a number of poems: Þormóðr Trefilsson's *Hrafnsmál*, *ÍF* iv. 67, 102, 124, 156, 168 (which like the verse in *Háttalykill* is not as regular as Snorri's example) and Sturla's *Hrafnsmál*, *Skj* B ii. 126–31, are the chief, but cf. also Sigurðr slembidjákn, *Hkr* iii. 312, the two quatrains in *Sturl.* ii. 211, Faulkes 1977–9, i. 388/37–40 and *Sǫrlastikki*, *Skj* A ii. 242.

Háttatal, stt. 80–94 are all examples of forms with end-rhyme (*runhent*). These are classified both according to the number of distinct rhymes in the stanza and according to the length of the lines (and their rhythm; cf. Kuhn 1983, 335). These vary from three to nine syllables and many of the patterns are similar to forms elsewhere in *Háttatal* apart from the presence of end-rhyme; presumably in principle Snorri believed that end-rhyme might replace internal rhyme in any metre (though the metre is rarely quite identical with the corresponding non-end-rhyming form; in particular *runhent* tends much more to end-stopping of lines). There are quite a lot of examples of *runhent* poems both before and after Snorri, but few of them correspond exactly to any one of his examples. Stanzas 82, 87, 88, and 90 have fairly near equivalents in *Háttalykill*, stt. 7 (there called *belgdrǫgur*), 24, 11, 17 (*rekit*; the stanzas in *Háttalykill* are similar in metre to those in *Háttatal*, though they do not always have the same rhyme-pattern). The best-known early example of *runhent* is Egill's *Hǫfuðlausn* (*ÍF* ii. 185–92), though if it is genuine the *lausavísa* 1 of Egill's father Skallagrímr (*ÍF* ii. 70) is the earliest extant *runhent* stanza. Both these examples use the short line (4–5 syllables, metrically equivalent to *fornyrðislag*) of *Háttatal*, stt. 80–1, 85–7; Egill sometimes uses masculine rhymes (like *Háttatal*, st. 87), sometimes feminine ones (like *Háttatal* 80–1, 85–6) and the rhymes are sometimes in pairs (*in minzta runhenda*), sometimes continue over four lines (*in minni runhenda*; he also sometimes has *skothending* between pairs of lines, a device not mentioned by Snorri; cf. *Háttalykill*, st. 17). The rhythms of *Háttatal*, stt. 80 and 86 appear in *Hǫfuðlausn*, stt. 5, 8, 10, 13, 14, and elsewhere. Skallagrímr's stanza is *in minzta runhenda* with masculine rhymes (*Háttatal*, st. 87; but rhyme is lacking in the final couplet). Modelled on *Hǫfuðlausn* seem to be Gunnlaugr's *Sigtryggs drápa* (*ÍF* iii. 75) and

Einarr Skúlason's *Runhenda* (*Skj* A i. 473–5; the latter uses only masculine rhymes, *Háttatal*, st. 87). *Fornyrðislag* with end-rhyme is also used by Bjǫrn Hítdœlakappi in *Grámagaflím*, *ÍF* iii. 168–9; by Þórðr Særeksson in his *lausavísa* 3, *Skáldskaparmál*, ch. 6; and in a stanza in *Friðþjófs saga*, *Skj* B ii. 293–4; but in most cases only masculine rhymes are used (*Háttatal*, st. 87), see Sighvatr 14 (*Skj* A i. 275, st. 2); Gunnlaugr ormstunga, *lausavísa* 2, *ÍF* iii. 69; Þjóðólfr Arnórsson's poem on Haraldr harðráði (*Skj* A i. 368); Sneglu-Halli, *lausavísa* 11, *Skj* A i. 390; Hólmgǫngu-Bersi, *ÍF* v. 76 and viii. 261; Magnús Þórðarson, *lausavísa* 1, *Skj* A i. 542; the anonymous verses in *Sturl.* i. 26, 519–20; Kolbeinn Tumason's religious poem, *Skj* A ii. 39–40, stt. 8–10; *Snjólfs vísur*, *Skj* A ii. 396–7. It is notable that this form is used both for *kviðlingar* and *flim* as well as for formal eulogy (and in religious verse). There are also examples, all with masculine rhymes only (*Háttatal*, st. 87), in *TGT* 125, 138, 147. The six-syllable *dróttkvætt*-type line of *Háttatal*, st. 88 is used by Gísli Súrsson, *lausavisa* 18, *ÍF* vi. 76; Óláfr Hávarðsson, *ÍF* vi. 297; Þorgils Hǫlluson, *ÍF* v. 194; Þórðr Rúfeyjaskáld, *Sturl.* i. 30 and in the anonymous st. 3 in *Sturl.* i. 21–2; Rǫgnvaldr jarl, *lausavísa* 31, *ÍF* xxxiv. 235; Hallr Þórarinsson and Ármóðr, *lausavísa* 3, *ÍF* xxxiv. 183, 212; Bjarni Kálfsson, *Skj* A i. 536; in the couplet in *Skáldskaparmál*, ch. 60, *SnE* (1931), 173/12; in the couplet in *Hkr* iii. 310; by Sturla Bárðarson, *lausavísa* 2, Guðmundr Ásbjarnarson, Þórir jǫkull, *Sturl.* i. 297–8, 398, 438 and in the anonymous stanza in *Sturl.* i. 391; by Guðmundr Galtason, *lausavísa* 2, *Skj* A ii. 44; in the anonymous quatrain in *ÍF* i. 110 n. 9; and in the Runic poem (*Skj* B ii. 248–9). The first two lines of Hjalti Skeggjason's *kviðlingr* (*Skj* A i. 139) are also like *Háttatal*, st. 88, though in some versions they are more like *Háttatal*, stt. 80–1 and 85 or 83, and the even line also has internal rhyme (cf. *dunhenda*, *Háttatal*, st. 24; the pattern is also found in Glúmr Geirason, *TGT* 94). It could be argued that this pattern of *runhenda* developed out of *dunhenda*. In *Harðar saga*, st. 18, in ch. 38 (1960, 173), there is a stanza with irregular six-syllable *runhent* combined with *dunhenda* (*Háttatal*, st. 24). Þorkell Gíslason's *Búadrápa*, *Skj* A i. 553–5, *Gnóðar-Ásmundardrápa*, *Skj* A i. 591, a stanza in *Sturl.* i. 251 and *Bárðar saga*, st. 3 (*Skj* B ii. 482) seem to be the only examples of the pattern of *Háttatal*, st. 83 (*málaháttr* with end-rhyme). *Málsháttakvæði* (*Skj* A ii. 130–6; included at the end of *Háttatal* in R) seems to be the first poem apart from *Háttalykill* to use the seven- to eight-syllable *runhent* of *Háttatal*, stt. 90–1 and 94. Unlike *Háttatal*, stt. 91 and 94 it often has eight syllables (sometimes including resolution) in odd lines and seven in even lines when the rhymes are masculine. There are further examples in *Sturl.* i. 257, *Skj* A ii. 463, st. 3 (like *Háttatal*, st. 91) and *Oddverja Annáll*, Storm (1888), 455, 457 (cf. JH–AH 130–1 and see *hrynhenda* above). *Háttatal*, st. 82 does not seem to be paralleled except in *Háttalykill*; there seem to be no parallels to stt. 84, 89, 92 (though st. 92 is not much different from st. 83; the rhythmical pattern occurs in *málaháttr*, st. 95) and 93 (unless perhaps in the *hálfhnept* stanza in *ÍF* xi. 331, which has end-rhyme in the last couplet).

Most of the above examples use *minzt runhenda* (lines rhyming in pairs); occasionally *minni runhenda* (rhyme continued over four lines) is found, but there seem to be no examples before Snorri's time of *full runhenda* (the same rhyme throughout the stanza). Snorri does not mention the possibility of the rhyme in one pair of lines (or half-stanza) forming *skothending* with that in another, though this feature is found in a number of stanzas in Egill's *Hǫfuðlausn* and occasionally elsewhere.

Bibliography

ALCUIN (1941). *The Rhetoric of Alcuin & Charlemagne*. Tr. W. S. Howell. Princeton, NJ.

ALDHELM (1919). *Aldhelmi Opera*. Ed. R. Ehwald. Berlin. Monumenta Germaniae historica. Auctores antiquissimi, 15.

AMORY, FREDERIC (1979). 'Tmesis in MLat., ON, and OIr Poetry: An Unwritten *notatio norrœna*.' *Arkiv för nordisk filologi*, 94: 42–9.

BEDE (1975). *Bedae Venerabilis Opera*, vi. *Opera didascalica*, i. Turnhout. Corpus Christianorum Series Latina, 123A.

Biskupa sögur (1858–78). Gefnar út af Hinu íslenzka bókmentafèlagi. 2 vols. Copenhagen.

BJARNI EINARSSON (1969). 'Andvaka'. *Afmælisrit Jóns Helgasonar. 30. júní 1969*. Reykjavík, 27–33. [Repr. Bjarni Einarsson (1987), 100–6.]

—— (1971). 'The Lovesick Skald'. *Mediaeval Scandinavia*, 4: 21–41.

—— (1987). *Mælt mál og forn fræði*. Reykjavík.

BJARNI GUÐNASON (1963). *Um Skjöldungasögu*. Reykjavík.

Bǫglunga sǫgur (1988). *Soga om Birkebeinar og Baglar*. Ed. Hallvard Magerøy. 2 vols. Oslo.

CHESNUTT, MICHAEL (1968). 'An Unsolved Problem in Old Norse–Icelandic Literary History'. *Mediaeval Scandinavia*, 1: 122–34.

Edda: Die Lieder des Codex Regius (1962). Ed. Hans Kuhn, vol. i. Heidelberg.

EINAR ÓLAFUR SVEINSSON (1958). *Dating the Icelandic Sagas*. London.

FARAL, E. (1924). *Les Arts poétiques du XIIe et du XIIIe siècle*. Paris.

FAULKES, ANTHONY (1977–9). *Two Versions of Snorra Edda from the 17th Century*. i. *Edda Magnúsar Ólafssonar*. ii. *Edda Islandorum*. Reykjavík.

FGT = *First Grammatical Treatise* (1972).

FIDJESTØL, BJARNE (1982). *Det norrøne fyrstediktet*. Øvre Ervik.

FINNUR JÓNSSON (1892). 'Navnet *Ljóðaháttr* og andre versarters navne, samt rettelser i texten i Codex regius af Snorres Edda'. *Arkiv för nordisk filologi*, 8: 307–22.

—— (1920–4). *Den oldnorske og oldislandske litteraturs historie*. 3 vols. Copenhagen.

—— (1933). 'Kenningers led-omstilling og tmesis'. *Arkiv för nordisk filologi*, 49: 1–23.

First Grammatical Treatise, The (1972). Ed. Hreinn Benediktsson. Reykjavík.

Flateyjarbók (1944–5). Ed. Sigurður Nordal. 4 vols. Reykjavík.

Flb = *Flateyjarbók* (1944–5).

FoGT = *Fourth Grammatical Treatise*. In *Den tredje og fjærde grammatiske Afhandling i Snorres Edda* (1884), 120–51.

Foote, P. G. (1982). 'Latin Rhetoric and Icelandic Poetry: Some Contacts'. *Saga och sed*, 107–27. [Repr. *Aurvandilstá* (Odense, 1984), 249–70.]

Fritzner, J. 1883–96. *Ordbog over det gamle norske Sprog*. 3 vols. Kristiania.

Gautreks saga (1900). Ed. W. Ranisch. Berlin.

Gordon, E. V. (1953) (Ed.). *Pearl*. Oxford.

GT Prologue = *Den tredje og fjærde grammatiske Afhandling i Snorres Edda* (1884), 152–5.

Gylfaginning: see Snorri Sturluson (1982).

Hákonar saga Hákonarsonar (1887). In *Icelandic Sagas*, ii. Ed. Gudbrand Vigfusson. London. Rolls Series.

Hákonar saga Hákonarsonar (1977). Ed. M. Mundt. Oslo. [The chapter numbers of the 1887 edition, which are those used here, are given in the margins.]

Hálfs saga ok Hálfsrekka (1981). Ed. Hubert Seelow. Reykjavík.

Hallberg, Peter (1975). *Old Icelandic Poetry. Eddic Lay and Skaldic Verse*. Tr. Paul Schach and Sonja Lindgrenson. Lincoln, Nebr. and London.

Halm, C. (1863). *Rhetores Latini minores*. Leipzig.

Harðar saga (1960). Ed. S. Hast. Copenhagen. Editiones Arnamagnæanæ, A6.

Háttalykill enn forni (1941). Ed. Jón Helgason and Anne Holtsmark. Copenhagen. Bibliotheca Arnamagnæana 1.

Helgi Sigurðsson (1891). *Safn til bragfræði íslenzkra rímna*. Reykjavík.

Helle, Knut (1958). *Omkring Bǫglungasǫgur*. Bergen. Universitetet i Bergen. Årbok. Historisk-antikvarisk rekke. Nr. 7.

Heusler, Andreas (1925). *Deutsche Versgeschichte*, i. Berlin und Leipzig.

—— (1950). *Altisländisches Elementarbuch*. Heidelberg.

Hkr = Snorri Sturluson (1941–51).

Hreinn Benediktsson (1968). 'Indirect Changes of Phonological Structure: Nordic Vowel Quantity'. *Acta Linguistica Hafniensia*, xi 1: 31–65.

Hrólfs saga kraka (1960). Ed. D. Slay. Copenhagen. Editiones Arnamagnæanæ, B1.

Hungrvaka (1905). In *Kristnisaga* etc. Ed. B. Kahle. Halle. Altnordische Saga-Bibliothek, 11.

ÍF = *Íslenzk fornrit* (Reykjavík, 1933–).

Jakob Benediktsson (1981). 'Hafgerðingadrápa'. In U. Dronke *et al.* (eds.), *Speculum norroenum*. Odense, 27–32.

JH–AH = Jón Helgason and Anne Holtsmark in *Háttalykill enn forni* (1941).

Jón Helgason (1936–8) (Ed.). *Íslenzk Miðaldakvæði*. 2 vols. Copenhagen.

—— (1953). 'Norges og Islands digtning'. In Sigurður Nordal (ed.), *Nordisk kultur*, viii: B. *Litteraturhistorie*, B. *Norge og Island*. Stockholm, 3–179.

JÓN ÞORKELSSON (1888). *Om digtningen på Island i det 15. og 16. Århundrede*. Copenhagen.

KEIL, H. (1855–80). *Grammatici Latini*, 7 vols. Leipzig.

KLNM = *Kulturhistorisk leksikon for nordisk middelalder* (1956–78).

KOCK, E. A. (1923–44). *Notationes norrœnæ*. 28 parts in 2 vols. Lund. [References are to paragraphs.]

—— (1933). 'Ett kapitel nordisk metrik och textkritik'. *Arkiv för nordisk filologi*, 49: 279–94.

KONRÁÐ GÍSLASON (1869). 'De ældste Runeindskrifters sproglige Stilling'. *Aarbøger*, 1869: 35–148.

—— (1875). 'Hljóðstafr, hljóðfyllandi (— hljóðfyllendr), hljóðfylling'. *Aarbøger*, 1875: 95–101.

Konungs skuggsjá (1920). [Ed. Finnur Jónsson.] Det kongelige nordiske Oldskriftselskab. Copenhagen.

Kristni saga (1905). Ed. B. Kahle. Halle a. d. S. Altnordische Saga-Bibliothek, 11.

KUHN, HANS (1937). 'Zum Vers- und Satzbau der Skalden'. *Zeitschrift für deutsches Altertum und deutsche Literatur*, 74: 49–63. [Repr. Kuhn (1969–78), i. 468–84.]

—— (1969*a*). 'Von Bragi bis Snorri'. *Einarsbók: Afmæliskveðja til Einars Ól. Sveinssonar 12. desember 1969*. Reykjavík, 211–32. [Repr. Kuhn (1969–78), iv. 115–32.]

—— (1969*b*). 'Die Dróttkvættverse des Typs "brestr erfiði Austra".' *Afmælisrit Jóns Helgasonar 30. júní 1969*. Reykjavík, 403–17. [Repr. Kuhn (1969–78), iv. 103–16.]

—— (1969*c*). 'Die Dróttkvættstrophe als Kunstwerk'. *Festschrift für Konstantin Reichardt*. Berne, 63–72.

—— (1969–78). *Kleine Schriften*, 4 vols. Berlin.

—— (1977). 'Hendingstudien'. *Sjötíu ritgerðir helgaðar Jakobi Benediktssyni 20. júlí 1977*. Reykjavík, 517–32.

—— (1981). 'Vor tausend Jahren: Zur Geschichte des skaldischen Innenreims.' In U. Dronke *et al.* (eds.), *Speculum norroenum*. Odense, 293–309.

—— (1983). *Das Dróttkvætt*. Heidelberg.

Kulturhistorisk leksikon for nordisk middelalder (1956–78). 22 vols. Copenhagen. [References are to columns.]

LP = Sveinbjörn Egilsson (1931).

MACRAE-GIBSON, O. D. (1989). 'Sagas, Snorri, and the Literary Criticism of scaldic verse'. In Rory McTurk and Andrew Wawn (eds.), *Úr Dölum til Dala: Guðbrandur Vigfússon Centenary Essays*. Leeds, 165–86. Leeds Texts and Monographs, NS 11.

MARTIN, ABBIE H. [1974]. *Snorri Sturluson's* Háttatal: *A Translation and Commentary*. Diss. Univ. of North Carolina at Chapel Hill (typewritten).
MEISSNER, RUDOLF (1921). *Die Kenningar der Skalden*. Bonn and Leipzig.
MÖBIUS, TH. (1879–81) (Ed.). *Háttatal Snorra Sturlusonar*. 2 vols. Halle.
MUNCH, P. A. (1857). *Det norske Folks Historie*, iii. Christiania.
MURPHY, G. (1961). *Early Irish Metrics*. Dublin.
Njála (1875–89). Det kongelige nordiske Oldskrift-Selskab. 2 vols. Copenhagen.
NN = Kock (1923–44).
NOREEN, A. (1923). *Altisländische und altnorwegische Grammatik*. Halle.
Norsk-islandske Skjaldedigtning, Den (1912–15). Ed. Finnur Jónsson. 4 vols. Copenhagen.
Nygaard, M. (1906). *Norrøn syntax*. Kristiania.
ÓH = *Saga Óláfs Konungs hins helga* (1941).
Óláfs saga Tryggvasonar en mesta (1958–61). Ed. Ólafur Halldórsson. 2 vols. Copenhagen. Editiones Arnamagnæanæ, A1–2.
ÓLAFUR HALLDÓRSSON (1969). 'Snjófríðar drápa'. *Afmælisrit Jóns Helgasonar 30. júní 1969*. Reykjavík, 147–59.
ÓTM = *Óláfs saga Tryggvasonar en mesta* (1958–61).
PAASCHE, F. (1957). *Norges og Islands litteratur inntil utgangen av middelalderen*. Oslo.
PATTERSON, W. F. (1935). *Three Centuries of French Poetic Theory*. 2 vols. Ann Arbor, Mich.
PE = *Poetic Edda*, cited from *Edda: Die Lieder des Codex Regius* (1962).
POOLE, R. (1982). 'Ormr Steinþórsson and the *Snjófríðardrápa*'. *Arkiv för nordisk filologi*, 97: 122–37.
QUINTILIAN (1920–2). *The Institutio Oratoria*. Tr. H. E. Butler. 4 vols. London. Loeb Classical Library.
Ragnars saga: in *Vǫlsunga saga ok Ragnars saga loðbrókar* (1906–8).
RASCHELLÀ, F. D. (1983). 'Die altisländische grammatische Literatur'. *Göttingische gelehrte Anzeigen*, 235: 271–315.
RAYNAUD DE LAGE, G. (1951). *Alain de Lille*. Montreal.
REICHARDT, K. (1969). 'A Contribution to the Interpretation of Skaldic Poetry: Tmesis'. In E. C. Polomé (ed.), *Old Norse Literature and Mythology: A Symposium*. Austin, Tex., 200–26.
Rhetorica ad Herennium (1954). Tr. H. Caplan. London. Loeb Classical Library.
RICHARDSON, PETER (1975). 'On the meaning of Old Icelandic *folk*'. *Semasia*, 2: 261–70.
Saga Óláfs konungs hins helga (1941). *Saga Óláfs konungs hins helga: Den store saga om Olav den hellige*. Ed. O. A. Johnsen and Jón Helgason. Oslo.
SAXO GRAMMATICUS (1979–80). *The History of the Danes*. Ed. H. E. Davidson, tr. P. Fisher. 2 vols. Cambridge.

SGT = *The So-Called Second Grammatical Treatise* (1982).

Sievers, E. (1878). 'Beiträge zur Skaldenmetrik'. *Beiträge zur Geschichte der deutschen Sprache und Literatur*, 5: 449–518.

—— (1893). *Altgermanische Metrik*. Halle.

Skáldskaparmál: in *SnE* (1848–87), i.

Skj = *Den norsk-islandske Skjaldedigtning* (1912–15).

Småstykker (1884–91). 16 pts. Udgivne af Samfund til Udgivelse af gammel nordisk Litteratur. Copenhagen.

SnE (1818): *Snorra-Edda*. Ed. R. Rask. Stockholm.

SnE (1848): *Edda Snorra Sturlusonar*. Ed. Sveinbjörn Egilsson. Reykjavík.

SnE (1848–87): *Edda Snorra Sturlusonar*. 3 vols. Copenhagen.

SnE (1924): *Edda Snorra Sturlusonar. Codex Wormianus. AM 242, fol.* Ed. Finnur Jónsson. Copenhagen.

SnE (1931): *Edda Snorra Sturlusonar*. Ed. Finnur Jónsson. Copenhagen.

SnE (1952): *Edda Snorra Sturlusonar*. Ed. Magnús Finnbogason. Reykjavík.

Snorri Sturluson (1941–51). *Heimskringla*. Ed. Bjarni Aðalbjarnarson. 3 vols. Reykjavík. [*ÍF* xxvi–xxviii.]

—— (1982). *Edda: Prologue and Gylfaginning*. Ed. Anthony Faulkes. Oxford. [Ch. nos. as in *SnE* (1848–87), i.]

So-Called Second Grammatical Treatise, The (1982). Ed. F. D. Raschellà. Florence.

Storm, G. (1888) (Ed.). *Islandske Annaler indtil 1578*. Christiania.

Sturl. = *Sturlunga saga* (1946).

Sturlunga saga (1946). Ed. Jón Jóhannesson, Magnús Finnbogason, Kristján Eldjárn. 2 vols. Reykjavík.

Sveinbjörn Egilsson (1931). *Lexicon poeticum antiquæ linguæ septentrionalis*. Rev. Finnur Jónsson. Copenhagen.

TGT = *Third Grammatical Treatise*. In *Den tredje og fjærde grammatiske Afhandling i Snorres Edda* (1884), 33–119.

Tredje og fjærde grammatiske Afhandling i Snorres Edda, Den (1884). Ed. Björn Magnússon Ólsen. Copenhagen.

Vésteinn Ólason (1969). 'Greppaminni'. *Afmælisrit Jóns Helgasonar 30. júní 1969*. Reykjavík, 198–205.

—— (1984). 'Sýnt í tvo heima'. *Pétursskip búið Peter Foote sextugum. 26. maí 1984*. Reykjavík, 58–60.

Vries, J. de (1964–7). *Altnordische Literaturgeschichte*. 2 vols. Berlin.

Vǫlsunga saga ok Ragnars saga loðbrókar (1906–8). Ed. Magnus Olsen. Copenhagen.

Wessén, E. (1915). 'Om kuida i namn på fornnordiska dikter'. *Edda*, 4: 127–41.

—— (1940) (Ed.). *Codex Regius of the Younger Edda*. Copenhagen. Corpus Codicum Islandicorum Medii Aevi, 14.

Glossary

The glossary is full but not complete: some ordinary words are omitted, and references are selective for words of frequent occurrence, but all technical words and all kennings in the verses are included with virtually complete references. 'n.' indicates that there is comment on the word in the notes. The following abbreviations are used:

a.	adjective	neg.	negative
abs(ol).	absolute(ly)	nom.	nominative
acc.	accusative	num.	numeral
adv.	adverb(ial)	obj.	object
art.	article	ord.	ordinal
aux.	auxiliary	o-self	oneself
comp.	comparative	p.	past
conj.	conjunction	part.	partitive
dat.	dative	pass.	passive
def.	definite	pers.	person
e-m	*einhverjum*	pl.	plural
e-n	*einnhvern*	poss.	possessive
e-s	*einhvers*	pp.	past participle
e-t	*eitthvat*	prep.	preposition(al)
e-u	*einhverju*	pres. (p.)	present (participle)
f.	feminine	pret.-pres.	preterite-present
gen.	genitive	pron.	pronoun
imp.	imperative	rel.	relative
impers.	impersonal	sg.	singular
indecl.	indeclinable	s-one	someone
inf.	infinitive	s-thing	something
instr.	instrumental	subj.	subject
interrog.	interrogative	subjunc.	subjunctive
intrans.	intransitive	subst.	substantive
irreg.	irregular	sup.	superlative
m.	masculine	sv.	strong verb
md.	middle voice	trans.	transitive
n.	neuter	vb.	verb
		wv.	weak verb

aðalhenda error for *alhenda* 42/11 textual n.

aðalhending *f.* chief rhyme, full rhyme 1/37, 8/18, 23/13, 35/9, 36/9, 37/9, 42/9, 44/9, 48/9–10, 54/10, 56/10, 58/12, 59/10, 63/11, 68/9–10, 71/10, 73/10, 75/11, 77/11, 78/9, 79/12.

aðalhendr *a.* having full rhyme 71/10.

áðr *adv.* previously; in the past 67/15.

afar *adv.* very 11/6, 46/4.

afhending *f.* use of the same rhyme-syllable in even lines as at the end of the preceding odd lines 47/10.
afkleyfisorð *n.* enclitic or proclitic (i.e. an unstressed grammatical word such as a pronoun, preposition, or conjunction) 73/10.
afkleyfissamstafa *f.* enclitic syllable, = *afkleyfisorð* 68/11–12.
afleiðing *f.* antecedent ('deducing') 15/10; dat. pl. 15/1 textual n.
ala (ól) *sv.* nourish (*e-u* with s-thing) 56/1.
álag *n.* extension, supplement 8/16.
álagsháttr *m.* 'extension-form' 27/1 textual n., 27/9. Cf. *TGT* 113.
ald- see **ǫld**.
alda *f.* wave; i.e. the sea 27/1; in kenning for gold, *logi ǫldu* 17/4; in kenning for mead, *hunangs ǫldur* 24/8 (obj. of *gefr*).
aldinn *a.* ancient 67/7.
aldr (rs) *m.* life 102/1; in kenning for head 50/8; = time? 94/4.
aldrlauss *a.* lifeless; sup., completely dead 51/4.
alframastr *a. sup.* quite the most outstanding; n. as adv. (predicative, see **vita**) 67/2.
alhenda *f.* 'complete rhyme' 43/10; *minni a.* 42/11, 43/9.
alhending *f.* 'complete rhyme' 44/11.
alhendr *a.* 'completely rhymed' 43/12 (n. as name of a verse-form).
alhneptr *a.* completely curtailed 77/13.
alinveldi *n.* 'ell-realm', forearm 43/8 (obj. of *hylr*).
allr *a.* all; n. as adv. all the way 1/5; pl. as subst., *allra* every one's 90/2.
allvaldr *m.* very mighty one; king 3/6; ruler, = Skúli 36/6 (subj. of *reyndisk*), 77/4 (subj. of *sér*, parallel to *ungr jarl*), 78/2 (subj. of *ferr*).
álmdrós *f.* 'bow-woman', valkyrie; in kenning for sword 60/1.
álmr *m.* (elm-)bow 31/3.
almætr *a.* entirely worthy 78/7.
alrauðr *a.* entirely red (of gold) 42/1.
alstýfðr *a.* (*pp.*) completely docked, i.e. having every line catalectic 51/11.
alvitr *a.* entirely wise, wise about everything; sup. most wise in every way 99/2.
ár (1) *f.* oar 22/6 (or **ár** n.?); in kenning for sword, *ár sára* 61/2.
ár (2) *n.* year 94/6; acc. of time 60/2.
árla *adv.* early (in the morning) 77/3.
árr *m.* messenger; in kenning for warrior, *ógnar árr* 62/1 (acc. pl.).
ásamt *adv.* together (*e-m* with s-one) 29/3.
askr *m.* ash (wood); spear 9/6, 57/3; in kenning for warrior (the poet), *askr þilju Hrungnis ilja* 30/3.
at (1) *prep.* (1) with dat. (2) with acc., after 23/4, 38/4. (3) elliptically with gen., at the dwelling of, in the presence of 41/7, 91/3.
at (2) *adv.* with comp., the 40/3, 7.

atriðsklauf *f.* 8/40: since the author gives neither definition nor example, it is difficult to know what he means by the term. The etymology (cf. *ríða*, *kljúfa*; in modern Icelandic *atriði* means 'item, event') suggests as possible meanings 'asyndeton' (*dissolutio*, *dialyton*), 'anacoluthon', or 'tmesis'. Cf. *TGT* 100.

átt *f.* family, clan 37/5 (gen. dependent on *þjóð*); *fram í átt* from generation to generation 89/4. Cf. **ætt**.

áttmæltr *a.* (*pp.*) containing eight separate utterances or sentences; n. as subst. as name of a verse-form 9/12, 10/1 textual n. Cf. **mál (2)**.

auðgjafi *m.* wealth-giver (i.e. the king) 13/3 (dat. with *lýtr*).

auðit *pp.* used impersonally with gen. and dat., granted; *e-s er a. e-m* s-thing is granted (by fate) to s-one 48/3.

auðkendr *a.* (*pp.*) easily recognized, easily picked out (of men's arms, because they have so much gold on them) 48/1.

auðmildr *a.* generous with wealth 11/2.

auðr *m.* wealth 27/7 (gen. dependent on *brjótr*), 37/6 (dat. with *hrauð*), 40/4, 42/1, 46/6 (gen. dependent on *ýtandi*), 48/1 (instr.), 48/4 (gen. with *auðit*), 89/3; in kenning for men 80/8.

auðsalr *m.* wealth-hall 102/2 (pl.).

auðspǫruðr *m.* one who is sparing of wealth, niggardly person 98/8 (parallel to *þá er vell gefa* 98/2).

auð-Týr *m.* wealth-Týr, generous prince, = Skúli 48/2.

auðviðr *m.* 'wealth-tree', man 48/3.

auka (eykr, jók) *sv.* increase, enlarge 6/19, extend 62/9 (*e-u* by s-thing); *fá e-t aukit e-m* cause s-thing to be increased for s-one 5/6; impers. *eykr e-u, e-u [er] aukit* s-thing is increased 7/12, 33/10, s-thing is added 34/9, 59/9; activate, bring into being: *a. Yggs feng*, *a. mærð* make poetry 31/3, 85/1. Cf. **stœra**.

austan *adv.*, *prep.* (1) with acc. *austan ver* east of the sea, i.e. in Scandinavia 82/6. (2) with gen. *austan fjarðar* east of the fjord or on the eastern side of the fjord (i.e. presumably the Vík (Oslo fjord), cf. *Hákonar saga*, ch. 74) 63/4 (with *kendi* or *stýrðu*?).

axl- *see* **ǫxl**.

bági *m.* enemy; in kenning for Óðinn, *úlfs b.* 3/1.

bál *n.* pyre; in kennings for gold, *ægis b.* 3/2 (cf. **ítr**), *Rínar b.* 91/6.

bálkr *m.* dividing wall; section (of a list), group 88/11, 91/10. In the name of the metre *Bálkar lag* 96/9, 97/12 it may have been a personal name. Cf. *TGT* 68.

bani *m.* death 64/6, 65/4 (gen. with *verðan*).

banna (að) *wv.* forbid (*e-t e-m* s-thing to s-one), prevent (s-one from using s-thing) 1/2, 37/1.

barð *n.* prow; pl. sides of prow (?) 74/7 (instr.); in kenning for ships 79/4; by synecdoche for ship (?) 3/3, 63/4 (obj. of *stýrðu*).

barr *a.* rough, violent 40/1.

bauggrimmr *a.* arm-ring-hating, i.e. generous 47/6 (with *baugstøkkvir*).

baugr *m.* (gold) arm-ring, bracelet 47/5 (instr.), 90/5.

baugstøkkvir *m.* arm-ring-flinger, i.e. generous ruler, = Skúli 47/5.

baugvǫllr *m.* plain of the (shield's) circle (i.e. of the disk round the shield-boss) = shield; in kenning for ruler 83/2.

beimar *m. pl.* men 47/2.

beita (tt) *wv.* with dat., use, brandish 98/3 (parallel to *gefa*).

bekkr (bekkjar, bekks) *m.* (1) stream 6/12; *sveita b.* i.e. bleeding wound 6/6; in kenning for ale, *blíðskálar b.* 87/3.

bekkr *m.* (2) bench; plank 76/1 (or by synecdoche = ship).

ben *f.* wound 11/7; in kennings for blood 56/2, 60/6 (with *legi*, see **lǫgr**).

benda (d) *wv.* bend (trans.) 20/5, 74/3; md. for pass. 9/5.

bera (bar) *sv.* carry 35/5; wear 44/5, 45/8; impers. *e-t berr* s-thing is carried, stretches 35/2; deliver, perform 68/8, pp. *borinn upp* lifted up, performed 97/4.

bergja (gð) *wv.* taste 11/7.

bíða (beið) *sv.* experience, gain 30/3.

biðja (bað) *sv.* bid, pray; with gen. demand, obtain 38/4 (inf. with *taka* or pres. pl.); with acc. and inf., pray that s-one may 30/1, with inf. understood, order s-one (to go) 44/15.

bifsœkir *m.* with gen., one who seeks to make s-thing quiver; *álms b.* warrior 31/3 (dat., i.e. for the warrior = Skúli).

bil *n.* hesitation 37/4.

bila (að) *wv.* fail, give way 8/5; perish, end 96/8.

bíta (beit) *sv.* bite, cut 10/7; abs., pierce 33/7.

bjartr *a.* bright, radiant 4/8 (glossed *glaðr* 4/17).

bjartveggjaðr *a.* (*pp.*) brightly sailed or bright-sided; sup. 34/4.

bjóða (bauð) *sv.* offer, command; teach, show (inf. after *gat*) 37/8.

bjoggju *see* **búa**.

bjórr *m.* beer 25/4.

bláferill *m.* dark path; in kenning for shield (part of kenning for warrior), *odds b.* 31/5.

blakkr *m.* (dark-coloured or pale? dun?) horse (used as a proper name in *Skáldskaparmál*, ch. 58); in kennings for ship, *b. brims* 35/5 (gen. pl. with *herfjǫlð*), *Haka b.* 38/7 (dat. of respect).

blár *a.* dark (in colour); of a sail 78/6; of spears 33/7; with *byrskíð* 74/5.

bláskíð *n.* dark ski; *barða b.* = ship 79/3 (instr., parallel to *randgarði*).

bleikr *a.* pale 27/2.

blíðr *a.* gentle, gracious 22/6, 79/3.

blíðskál *f.* pleasant bowl, cup that cheers; in kenning for ale, *blíðskálar bekkr* 87/4.
blik *n.* gleam; in kennings for gold, *b. spannar* 40/8, *b. brimlands* 45/5.
blikna (að) *wv.* gleam 45/5.
blikurmaðr *m.* miser 40/6.
blóð *n.* blood 6/12, 18/1, 14, 51/8, 56/1 (dat. with *byrsta*, or parallel to *bens rauðsylgjum*; cf. **byrstr**).
blóðdrykkr *m.* blood-drink 11/8.
blóðsvǫrr *m.* blood-bird (*svǫrr*: a species of bird), i.e. eagle or raven, in kenning for warrior 92/8.
bogi *m.* bow; in kenning for arms, *boga nauðir* 48/2; in kenning for hail of arrows 62/6.
bolr *m.* trunk, body 10/4.
borðgrund *f.* 'board-ground', i.e. sea 74/3.
borg *f.* fortification, castle; *vilja b.* = breast 51/5.
bót *f.* improvement 16/17.
bráðr *a.* quick; n. as adv. swiftly, hard 78/5.
bragarbót *f.* 'poem's (poetry's) improvement', name of a verse-form 31/9.
bragarháttr *m.* 'poetic metre', name of a verse-form 31/1 textual n.
bragarmál *n.* 'speech of poetry, poetic speech', i.e. elision (of the vowel of an enclitic) 8/20. Cf. *TGT* 87.
bragnar *m. pl.* men; gen. with *bǫl* 24/3, *þǫgn* 25/3, *sagna* 59/8, *ótali* 88/4, *vinr* 90/6; dat. with *þekkr* 87/3.
bragningr *m.* prince; = Hákon 24/4 (subj. of *gleðr*); = Ingi Bárðarson 34/3; = Earl Skúli 36/8 (acc. with inf. after *frá*), 45/2; pl. = Skúli and Hákon 96/7.
bragr *m.* poetry, poem; poetic form 100/2; *bragar lag* verse-form 70/2.
braka (að) *wv.* crack, make a cracking sound 9/8.
brandr (1) *m.* firebrand 18/12–13; in kenning for gold (arm-rings), *vallands branda* 44/8 (with *svala* [*refhvarf*], obj. of *bera*).
brandr (2) *m.* sword 8/8, 9/7, 14/3 (instr.), 18/8, 19, 98/3; in kenning for shield 59/4.
brattr *a.* steep 3/3.
bregða (brá) *sv.* move; *b. á e-t* act in accordance with s-thing 26/2; md. be altered, be affected, cease to be 28/13.
breiðfeldr *a.* extensive 14/4 (with *lǫnd*).
breiðr *a.* broad 36/7 (with *fylking*).
breki *m.* wave, breaker 77/4; pl., obj. of *lemr* 38/2.
bresta (brast) *sv.* break (intrans.); be broken off (to be given as a gift) 49/2, 90/5. Cf. **brjóta, brjótr**.
breyta (tt) *wv.* with dat., vary, change (cf. **skipta**; usually impers. pass.) 1/44, 52, 6/16, 8/45, 9/11, 27/12, 14; *breytt til* changed to? in accordance with? modelled on? 40/9; pp. as a. 0/11, 11/10.

brim *n.* surf; *b. horna* = beer 25/3; *blakkr brims* = ship 35/6.
brimdýr *n.* 'surf-animal', i.e. ship 74/4.
brimland *n.* 'surf-land', i.e. sea; in kenning for gold, *brimlands blik* 45/6.
brjóst *n.* breast 6/13 (pl.).
brjóta (braut) *sv.* break (trans.) 22/4, 59/3, 61/5, 74/5, 78/3, 93/5; i.e. distribute (cf. **bresta**) 26/1, 89/7.
brjótr *m.* breaker, distributor; *b. auðs* i.e. generous man (Snorri) 27/5.
bróðir *m.* brother 22/10 (metaphorical); *hilmis b.*, *bragnings b.*, *konungs b.* = Skúli, (half-)brother of King Ingi Bárðarson 33/5, 34/3, 69/2 (dat. of advantage).
brot *n.* fragment 45/1 (pl.).
brotna (að) *wv.* break in pieces, fragment (intrans.) 45/1 (to make rings fragment means to distribute gold).
brott *adv.* away 47/4.
brún *f.* brow; edge (i.e. shore) 3/4.
bruna (að) *wv.* advance with speed 72/6.
brúnn *a.* brown, burnished (of a sword) 50/4 (with *glóð*).
brynja *f.* mail-coat 9/7, 57/4; *brynju él* storm of missiles 62/4.
búa (bjó) *sv.* prepare (*e-m* for s-one) 11/2, 49/5; fit out 98/4; *b. e-t e-u* provide s-thing with s-thing (man a ship with a crew) 34/1; *b. e-t e-m* (*e-u*) cover, adorn s-thing (with s-thing) for s-one 32/5, 48/5; p. inf. *bjoggju* made ready 36/7 (acc. (*bragning*) and inf. after *frá*); compose 69/1; *búa of* be endowed with 15/6; pp. *búinn til* ready for 88/5.
búandmaðr *m.* peasant 33/8; gen. with *støkr* 35/8.
buðlungr *m.* prince 15/8; = Hákon 14/3, = Skúli 74/8 (subj. of *veit*).
byrja (jað) *wv.* begin, undertake 38/1.
byrr *m.* (following) wind 38/6, 78/5; = storm, in kenning for battle, *brands stranda b.* 59/3.
byrskíð *n.* wind-spar or -ski, i.e. ship 74/6 (pl., acc. with *veit*).
byrstr *a.* (*pp.*) bristled (*e-u* with s-thing) 56/1 ('bloody-bristled', i.e. with blood on its bristles or having gouts of blood looking like bristles (see Fidjestøl 1982, 73; *NN* 3261); with *ylgi*).
bœr *m.* dwelling; *heila b.* = skull 2/6.
bǫðharðr *a.* battle-hard 74/7 (with *buðlungr*).
bǫl *n.* evil, trouble, misery (malice?) 24/3.
bǫrr *m.* a species of tree (spruce?); in kenning for man, *auðs b.* 80/8.

dáð *f.* deed, ability to achieve great deeds, valour 93/2.
dáðrakkr *a.* mighty in deeds, valiant 5/3 (with *dreng*).
dagr *m.* day; acc. of time 56/4.
darraðr *m.* spear; (collective) gen. with *skúrum* or *skǫpt* 52/2.
dauði *m.* death 5/3.
deila *f.* dispute, strife 19/8 (obj. of *skapar*), 19/12.

deilir *m.* distributor; in kennings for ruler (Hákon), *d. gulls* 2/6, *d. styrjar* 21/7, 28/7.

detthendr *a.* 'falling rhymed' 8/16, 28/14; n. as subst. 29/1 textual n. (the name presumably refers to the 'falling' rhythm in even lines [ˊ ˋ ×], cf. JH–AH 131).

djúpr *a.* deep 19/11; with *skip*, high? low in water?—or with *fǫll*? (see **grunn**) 19/4.

dólgr *m.* enemy 66/4 (with *of*); in kenning for ruler (i.e. Hákon) 30/8, for Þórr 3/8.

drag *n.* an extra line at the end of a stanza (?) 8/31 n.

draga (dró) *sv.* drag 44/17; draw, pull 8/7; (metaphorically) 16/10; stretch (in meaning)? use (*til* for)? 16/14; *dregr þat til at* this results in this, that 8/13 n.; md. for pass. 16/3; *dragask fram eða aptr* be drawn forward or back, be increased or diminished 6/23; *dragask eptir* be similar to, be reminiscent of 39/10. In *SGT* 66 *dreginn* is used to mean long (of a vowel).

dragreip *n.* halyard 77/8 (subj. of *mœta*).

drápa *f.* a formally constructed poem (usually distinguished from a *flokkr* by having *stef* or refrains) 35/13.

draughendr *a.* 'ghost-rhymed' or 'trunk-rhymed'? 64/9.

draugr *m.* trunk (of tree); in kenning for men, *hjarar d.* 47/6.

draugsháttr *m.* 'ghost-form' or 'trunk-form'? 29/10, 30/1 textual n.

dreifa (ð) *wv.* scatter 17/19.

drekka (drakk) *sv.* drink 87/2, 91/3; *d.* (*menn*) *glaða* 'drink men glad', give men drink so that they become happy 86/5 (obj. is *vandbaugskaða*).

drengr *m.* man, warrior 4/6, 14/2; collective sg. 5/4.

dreyrfár *a.* blood-coloured, blood-stained 7/6.

drífa (dreif) *sv.* drive; intrans. of ships 19/2, of rings (i.e. they move quickly, are freely given) 87/1; with dat. cause (missiles) to be driven 62/3 (obj. is *brynju éli*); impers. with dat. s-thing is driven, scattered ('it snows with') 42/1.

dript *f.* snow-fall; *boga d.* = hail of arrows 62/6 (obj. of *heldr*).

drós *f.* lady; *Yggs d.*, i.e. a valkyrie; perhaps = Hildr, i.e. battle, in kenning for sword 50/1.

drótt *f.* court, following of a king or earl 15/3, 85/6, 93/7.

dróttkveðinn *a.* (*pp.*) composed in *dróttkvætt* 6/17 n.

dróttkvæðr *a.* pertaining to court poetry; in accordance with *dróttkvætt* 62/12, 65/14; *dróttkvæðr háttr* the verse-form known as *dróttkvætt* 1/26, 37, 41–2, 7/10, 8/11, 15, 25, 9/11 etc.; pl. 23/11, 53/13; *dróttkvæð hrynjandi*: *hrynjandi* based on *dróttkvætt* 62/15; n. as subst. *dróttkvætt* 9/10, 34/11, 35/10, 54/12, 59/10, 62/10, 68/10, 69/11, 75/12, 77/12, 82/10, 88/10, 97/12.

drǫgur *f. pl.* 'drawings' (cf. **draga, drag**) 16/1 textual n., 16/10.

duga (ð) *wv.* (1) help 18/13. (2) abs., be capable, act well, be valiant 8/2, 18/8, 20; impers. with (*at* and) inf. it is a good thing, it is worth (doing s-thing) 27/6, 81/7.

dul *f.* secrecy, concealment 18/2, 9, 10, 15.

duna *f.* noise, din; in kenning for battle, *d. geira* 53/2.

dunhenda *f.* 'echoing rhyme' 24/1 textual n.

dunhendr *a.* 'echoing rhymed' 8/16 n., 47/10–11; n. as subst. 23/14 textual n., 24/1 textual n. Cf. *FoGT* 147.

dvala (ð) *wv.* delay, hold back (trans.) 68/2.

dvǫl *f.* stay, rest 20/3; staying (in a place) 90/2 (pl.).

dynblakkr *m.* noisy horse; in kenning for ship, *stáls d.* 31/1. Blakkr is the name of a mythical horse in *Skáldskaparmál*, ch. 58.

dynbrími *m.* clashing flame; in kenning for sword, *hræs d.* (or *hræs glæs d.*) 50/6.

dynbrunnr *m.* rushing (resounding) stream; in kenning for blood, *hræs d.* 32/4.

dynr *m.* noise; in kenning for battle, *málmskúrar d.* 39/4.

dýr *n.* animal; in kenning for ship, *unna d.* 28/6 (pl.).

dýrð *f.* glory 85/5 (obj. of *segja*).

dýrr *a.* dear, valued, precious 82/8 (sup. with *jarla*, = Skúli, and predicative); sup. with *myklu* 94/8; of a verse-form, elaborate, ornate, splendid 36/10.

døkkr *a.* dark (in colour); of ships 73/2 n. (with *hreina*).

eða *conj.* or 55/3, 96/8 (links *farisk* and *bili*).

efla (d) *wv.* make strong, increase, enact; wage (war) 94/3. Cf. **stœra**.

efni *n.* material, speech-content 4/9.

efri *a. comp.* latter 19/11.

egg *f.* edge (usually by synecdoche = sword) 4/2, 12 (subj.), 5/2, 9, 10, 8/2, 10/7, 32/3, 33/6 (subj. of *hrauð*); instr. 56/6 (with *nýbitnar*; perhaps to be taken as forming a compound word with it, see *NN* 1318); instr. (with *bjoggju*), or possibly as first part of compound by tmesis with *-þing* 36/7.

egna (d) *wv.* use s-thing (*e-u*) as bait; entice, provoke (with s-thing); *e. til ógnar tognu sverði* aim to bring about battle with drawn sword 58/4.

eiða *f.* mother (see *Skáldskaparmál*, ch. 68); *orms váða e.* = Jǫrð, i.e. the land (of Norway) 3/5.

eiga (á, átta) *pret.-pres. vb.* have 7/5, 26/7; *átti ráða* had to rule (over: with *þeim er*) 15/3 (subj. is *hans faðir*); get 40/7.

eiginn *a.* own; his own 14/7 (with *ríki*).

eignask (að) *wv. md.* absol., gain possessions 9/3.

eik *f.* oak; by synecdoche = ship 27/2 (instr.), 72/5.

einn *pron. a.* same 1/30, 31, 35, 8/26, 16/12, 80/9; only 8/45, 9/11.

einnig *adv.* = *einn veg*, in the same way 36/10.

einstaka *a. indecl.* single, not forming part of a series; *e. vísa* = *laus vísa, staka* 8/27.

eisa *f.* ember, fire; in kennings for gold, *e. lýsheims* (of gold as ornament on a successful king's person, or on his followers) 22/8 (subj. of *náir*), *e. fens* 26/1; in kenning for sword, *Yggs drósar e.* (i.e. fire of valkyrie or of battle) 50/1.

él *n.* storm, in kennings for battle (hail of missiles): *lindar snarvinda skýs él* 32/1, *brynju él* 62/4 (obj. of *drífr*).

eldr *m.* fire 17/10; in kenning for sword, *Hlakkar e.* 57/5, for gold (in kenning for ruler), *e. lagar* 69/7.

eljunstrǫnd *f.* 'energy-shore', home of determination, i.e. breast 63/5 (obj. of *rendi*).

elli *f.* old age 61/7 (acc. with *þrotna*, or with *spyr*); *til e.* until (your) old age 3/8, 30/8.

elligar *adv.* otherwise 15/11.

élsnœrðr *a.* (*pp.*) storm-bound 79/7.

en *conj.* but, and, and moreover 15/7, 16/2, etc.; with comp., than 91/8 (with *œðra*), 98/6, 8 (with *verðari hæra hróðrar*; *en heimdrega* is parallel to 98/3–4, *en auðspǫruð* to 98/2); with *fyrr* 102/8.

endask (nd, að) *wv. md.* come to an end, conclude 71/11. Cf. **lúka(sk)**.

endi *m.* end (of the line) 32/10, 37/10.

eptir *prep.* with acc., at (after) the end of 59/10; as adv., behind; *er e. sem* what remains is as 62/10.

ern *a.* keen, energetic, able (with gen. = with s-thing) 79/5 (with *jarl*).

eyða (dd) *wv.* with dat., destroy 5/1, 65/7, 66/1.

eyðir *m.* destroyer; in kenning for battle-leader, ruler, *e. baugvalla*, = Skúli 83/2 (vocative, but see n.).

eyrindi *n.* stanza, strophe (= *vísa*; cf. **ørindi**) 0/21, 1/10.

fá (fekk, pp. **fengit)** *sv.* get, receive, obtain 14/2, 39/7, 84/1; *fá e-m e-t, fá e-m e-s* provide s-one with s-thing 11/5, 13/1; with pp., be able to 21/5, 51/7, 100/6, cause (s-thing to be s-thing) 5/5; md., be gained 9/3.

fagna (að) *wv.* with dat., welcome, rejoice in 14/7; *f. við* rejoice at, receive joyfully 88/3.

fagr (rt) *a.* fair, fine 91/4.

fagrdrasill *m.* fair steed; in kenning for ship, *f. lǫgstíga* 22/4.

fagrregn *n.* fair rain (i.e. tears), in kenning for gold: *f. Mardallar hvarma* 42/6.

fagrskjaldaðastr *a.* (*pp.*) *sup.* provided with (having) the most beautiful shields 34/2.

falbroddr *m.* (spear-)socket-point; gen. pl. with *ern*, eager to use spears, skilful with spears (or perhaps with *oddum*, parallel with *hrænaðra*) 79/6 textual n.

falda (felt, *pp.* **faldinn)** *sv.* dress the head; *faldinn grímu* crowned with a helmet 15/2; *f. rauðu* dress one's head in red (a red cap), i.e. get a bloody head, be wounded 63/2 (with *kendi*).

fall *n.* fall (of water), breaking water, breakers 19/3 (cf. **grunn**), 77/4 (obj. of *sér*); metaphorically, case? *í eitt fall* at the same time or with the same phrase 17/13 n.; as a metrical term, *tíðar fall* cadence? 16/13 n. Cf. *GT Prologue* 154, 155: *fall ęðr tima, einnar tiðar fall*, where the meaning seems to be 'quantity'. In *TGT* and *FoGT*, *fall* means 'grammatical case'.

falla (fell) *sv.* fall 2/1, 20/2, 42/5, 61/1; in battle 10/5, 53/7; subjunc. (optative) let it fall 102/5 (cf. **lof**); be fitting 75/3 (subj. *mál*); pp. *vel fallinn* very fitting 67/13 (cf. *FoGT* 136); occur 65/12.

falr (1) *m.* socket of spear; by synecdoche = spear 9/4, 18/5, 18.

falr (2) *a.* (freely) available 28/6, 43/8 (with *meldr*).

fang *n.* tunic (of a woman); *Hamðis fang* coat of mail 2/3 (subj. of *fellr*; cf. textual n.).

far *n.* vessel 22/8.

fara (fór) *sv.* go 6/3, 90/1; travel 17/13, 78/1; be arranged 2/10, 62/14, 63/10, 69/9; impers. pass. with dat., s-thing is arranged 57/9; with inf., go and, go to 6/12, 55/8; *f. eptir* follow 1/36, 13/11, imitate 4/20; *f. saman* be consistent 1/27; md., perish 96/6.

fegra (rð) *wv.* make (more) elegant, adorn 4/19 (absol.), 6/20.

fegri *a. comp.* (of **fagr**) more elegant 1/21; sup. 44/10. Cf. *FoGT* 137.

feigr *a.* doomed, about to die 65/2.

fela (fal) *sv.* cover 2/5 (subj. is *skǫrungr*, unless it is understood from the previous clause).

fella (d) *wv.* cause to fall, put an end to, dispel 25/3.

fellir *m.* feller; in kenning for war-leader, ruler (= Hákon), *dólga f.* 30/8 (vocative, parallel to *vígfoldar valdi* unless it belongs with *nýtr vartu oss*); = sword in kenning for blood, *hræs f.* 60/4 n.

fémildr *a.* generous with money 29/4.

fen *n.* fen, lying water; in kenning for gold 26/2.

fengr *m.* booty; in kenning for poetry, *Yggs f.* 31/4 (cf. *Skáldskaparmál*, *SnE* 1848–87, i. 224).

fenna (nt) *wv. impers.* cover (as with snow); pp. with dat., covered with s-thing (as with snow) 65/8 (with *fjǫrnis hlíð*, i.e. head).

ferð (1) *f.* movement, travel, journey 13/7, 20/3 (word-play with **ferð (2)**).

ferð (2) *f.* troop, crew, usually = men 8/4 (gen. with *sverð*), 20/3, 37/3 (subj. of *vann*), 54/6; *gotna f.* 91/2, *ýta f.* (gen. with *eljunstrandir*) 63/6; in pl. = men, soldiers 33/2.

ferri (= **fjarri**) *adv.* far (with dat., from being like s-thing) 40/2.

festa (st) *wv.* fasten; ensure, guarantee (*e-m* to s-one) 37/4 (pp. with *vann*); betroth o-self to (*f. Hildi*: metaphor for engage in battle) 49/2.

festr *f.* cable, hawser; in kenning for ship, *hestr festa* 71/6.

féstríðir *m.* money-tormentor, enemy of money, i.e. generous ruler (Skúli) 43/5.

fet *n.* tread, step 32/7 (variants *fit* TW, *fót* U).

fetilhams *m.* baldric-slough, kenning for sword-fittings or scabbard 6/4.

fetill *m.* baldric, strap for hanging the sword from the shoulder 6/10 (pl. of the fittings collectively).

finna (fann) *sv.* find 16/14, 28/5; invent 35/10 (*fann fyrst* was the first to invent); with suffixed neg. *finnrat* cannot find, will not find 92/5; md., be found, appear, occur 0/19, 1/14 (with inf.), 6/23, 8/30, 16/15, 44/11 (*í* in it), 53/13; be considered 82/10.

firar *m. pl.* men 13/2; dat. of respect (advantage) 42/2 (with *á vals reitu*, i.e. onto men's arms), 50/4 (with *í hof hugtúns*).

firrask (ð) *wv. md.* be separated from (be put aside by) 23/1; be deprived of (or shun?) 20/3 ('get no rest').

first *adv. sup.* (of **fjarri, ferri**) farthest off, farthest apart 28/12, 31/10; with dat. furthest from (being) 40/6.

fit *f.* limb 42/5 (*þegnum* is dat. of respect or advantage); of an animal: paw 56/7. Cf. **fet**.

fjall *n.* mountain 13/4; in kenning for waves, *svana fjǫll* 76/5 (acc. with *hleypir*).

fjallvargr *m.* mountain-wolf 53/8.

fjarri (= **ferri**) *adv.* far, a long way 35/2.

fjórði *ord. num.* fourth 69/5 (with *lofun*).

fjórðungalok *n.* (*pl.* ?) 'quarter-ends', name of a verse-form 11/1 textual n.

fjórðungr *m.* quarter (stanza), i.e. a couplet 1/10, 11, 22, 18/16, 17, 19; *f. vísu* 1/20, 26.

fjǫlmennr *a.* having a great company or following 29/8 (with *stilli*); predicative (vb. to be understood) 69/4.

fjǫlsnœrðr *a.* many-stranded, complicated 68/4.

fjǫlvinjaðr *a.* having many meadows, i.e. resting-places, places on which ornaments may be put 43/7 (with *alinveldi*; if one adopts the reading of R*, *-vinjaðr*, the word belongs with *meldr*, and must mean 'which finds many resting-places, i.e. recipients').

fjǫlvitr *a.* wise about many things, in many ways, most wise 55/2 (with *œgi*).

fjǫr *n.* life 62/4.

fjǫrðr *m.* fjord 63/4 (i.e. the Vík? see **austan**).

fjǫrnir *m.* 'life-protector', helmet; in kennings for head 62/5, 65/6.

fjǫrvar *m. pl.* men 80/1 (with *fyrir*).

flagðaháttr *m.* 'ogre-form' 33/11.

flagðalag *n.* = *flagðaháttr* 34/1 textual n.

flatr *a.* flat (of rings!) 90/5.

flaust *n.* a kind of ship (etymologically related to *fljóta* sv.) 19/6 (indirect obj. of *skapar*), 23/3 (pl.), 34/1.

fleinbrak *n.* 'spear-noise', (turmoil of) battle 2/2, 2/12 ('to call battle *fleinbrak*').

fleinn *m.* shaft, spear; in kenning for shield 65/5.

fleinstýrir *m.* spear-guider, = Snorri 29/6 (dat.).

fleinþollr *m.* spear-fir, warrior, = Skúli 75/3 (subj. of *lætr*).

flestr *a. sup.* most, nearly every 16/16, 49/6 (dat. sg.), 60/2 (with *ár*); i.e. more than anyone else 93/3.

fljóta (flaut) *sv.* float 44/18.

fljótr *a.* quick 17/18; n. as adv. hastily 17/5 (with *válkat*), 17/17, 17/30.

flytja (flutta) *wv.* cause to move (*braut* away) 17/15, 23; bring forward, present, deliver, perform 80/1 (pp.); with suffixed pron. *-k* 81/1, 95/3 (*e-m* to, for s-one). (Since Snorri probably did not actually perform the poem to its recipients, having composed it in Iceland after his return from Norway, the word can perhaps be taken to mean simply 'compose' here; cf. Möbius 1897–81, i. 34.)

fold *f.* land; pl., obj. of *varða* 18/3, 55/7; = Norway 39/5, 53/4 (obj. of *halda*); = the earth 102/6 (subj. of *falli*); in kenning for head, *fjǫrnis f.* 62/5.

fólk *n.* (1) people, army, host 10/5, 34/1 (perhaps **fólk** (3)?), 36/3, 55/7, 62/6 n. (2) = battle 37/3 (with *herði*). (3) ? sword 60/8 (pl., obj. of *fylla*); as first element of compound separated by tmesis, *fólkskúrum* 'sword-showers' (or 'battle-showers', 'army-showers'), i.e. rain of weapons, warfare? 62/6 n. See 60/8 n.

fólkglaðr *a.* battle-glad (people-glad?) 69/6 (with *friðrofa*).

fólkhamla *f.* host-rod, kenning for sword 24/5 (gen. with *viðum*).

forn *a.* old (of beer) 25/4; ancient, from old times 13/12, 34/12, 53/13.

fornafn *n.* pronoun 1/23. In *Skáldskaparmál*, chs. 1 and 67 the word is used of a kind of kenning or *heiti* that replaces a proper name.

fornkvæði *n. pl.* ancient ('classical') poems 58/16 (cf. 53/13).

fornskáld *n.* ancient poet, classical (Norse) poet 8/30, 53/11 and textual n., 58/14 (cf. *FoGT* 136, 147, *GT Prologue* 152–3).

fornyrðislag *n.* 'old-story metre' (or 'old-talk metre'?) 95/9, 97/9.

fram *adv.* forward 20/4, 33/4, 64/2, 69/5; in front 56/8 n.; on one's face 32/8; on (in time) 89/4; *um fram* in addition 67/11.

framan *adv.* in front, at the beginning 62/9.

framar *adv. comp.* further forward 18/19.

framast *adv. sup.* to the highest degree (i.e. the superlative) 34/10.

frami *m.* advancement, benefit 17/25 (cf. **framr** and **fremja**); honour, glory 44/1 (obj. of *skotnar*), 67/8 (*e-m* for s-one), 81/2 (gen. with *græði*), 88/8, 93/8 (gen. with *gnótt*).

framla *adv*. excellently, honourably, gloriously 4/4; in great measure, generously 24/5.

framligr *a*. excellent, outstanding 4/14.

framlyndr *a*. bold in nature, bold-hearted 24/6 (with *skjǫldungr*).

framr *a*. forward(-thrusting), outstanding, bold 4/2, 12, 17/8, 24 (cf. **fremja**), 32, 72/4 (with *seggir*), 94/2 (with *Kraki*).

frár *a*. swift, keen 60/7, 75/4 (with *fleinþollr*).

fregna (frá) *sv*. hear, learn, with acc. and inf.; *frá ek lung geisa en herða svǫrð* 21/1, *frá ek bragning bjoggju* 36/7, *frágum Haka efla* we have heard that H. waged 94/3; with acc. *fregnum styr hans* 62/8.

frekr *a*. bold, strong, violent 38/2.

frelsa (t) *wv*. free 1/3.

fremð *f*. honour, advancement 101/1.

fremja (framða) *wv*. move forward (trans.), advance, promote; literally 18/6 (pp.), 12, 18, 19; metaphorically, benefit 18/1 (i.e. battle takes place, providing drink for the wolf), 18/14; *f. e-n e-u* benefit or honour s-one with s-thing 43/1, 46/5, 47/5, 83/7; at 17/24 *frǫmum* is taken as 1st pers. pl. of *fremja* (in the verse as dat. pl. of *framr*); perform, wage 10/6, 40/1, 58/8; md. (for pass.), be performed 36/3.

fremri *a. comp*. (of **framr**) more outstanding, superior; *at fremri* (so much) the greater 1/7; sup. *fremstr* most outstanding 68/1 (with part. gen.: sense continued at 70/8).

friðask (að) *wv. md*. be (left) in peace, be undisturbed 44/7 (subj. *herfjǫlð*; or taking the subj. as *valdi, friðask við* 'make peace with'; in either case the phrase means that the gold is always being given away by Skúli. See *NN* 3146).

friðbygg *n*. 'peace-barley', in kenning for gold (the peace-giving grain of Fróði's mill), *Fróða f*. 43/6.

friðlæ *n*. peace-destruction, i.e. warfare 17/6, 20.

friðr *m*. peace 17/20, 19/8 (obj. of *viðr*), 19/12, 33/6.

friðrof *n*. breaking of peace, hostility 1/4 (gen. with *ofsa*).

friðrofi *m*. breaker of peace, in kenning for ruler: *f. elds lagar* he who does not leave gold in peace, generous giver (Skúli) 69/8 (gen. with *lofun*).

frost *n*. frost; in kenning for sword (or battle?—cf. **hrími**), *Mistar f*. 61/3.

frumhending *f*. (= *fyrri hending*) anterior rhyme, the first of the two rhyme syllables in a line of *dróttkvætt* verse 1/39, 8/13, 28/12.

frumsmíð *f*. first attempt (at making s-thing) 16/16.

frœði *n. pl*. information, knowledge, accounts of deeds 81/1.

frœkn *a*. bold, courageous 4/17, 52/4; comp. (i.e. more courageous than Skúli) 92/5.

fullframr *a*. very bold 84/3 (with *fylkir*).

fullhvatr *a.* very bold, vigorous 5/2, 10.
fullkominn *a.* (*pp.*) complete 9/9 (cf. **fylla**).
fullna (að) *wv.* extend, amplify 34/10.
fullr *a.* full 23/8 (with *skál*), 43/10, 44/11 (i.e. proper), 83/9, 92/9; complete 13/10, 27/10, 13, 47/8.
fullseðjask (-sadda-) *wv. md.* with gen., *f. hungrs* fully sate one's hunger 51/2.
fullsterkr *a.* most mighty 36/4 (with *verk*).
fundr *m.* meeting (*e-s* with s-one) 101/2; *til e-s fundar* to meet with s-one 27/3.
furask (að) *wv. md.* (vowel length uncertain) be furrowed, damaged 4/7, 17.
fúrr *m.* fire; in kenning for sword, *fleinbraks f.* 2/1 (pl.), 13.
fúss *a.* eager (used as equivalent of adv.) 26/1, 29/5.
fylgja (lgð) *wv.* with dat. accompany, go with 5/9, 11, 8/31, 38/10, 61/9, 85/10; follow 63/10; *f. fyrir* precede 36/10 n.; belong to 15/10, characterize, be characteristic of 67/12, 93/2 ('he possesses'); *er þeim fylgja* of that kind (class) 44/12; be contained in 1/11 ('each line consists of six syllables'), 1/31; comprise, make up 62/12; be involved in 1/25.
fylki *n.* district, county 17/29.
fylking *f.* battle-line 36/8 (obj. of *bjoggju*).
fylkir *m.* war-leader, ruler; = Hákon 9/3, 17/5, 32, 26/1, 28/5, 29/5; = Skúli 38/1, 65/5, 84/4. Cf. **fólk**.
fylkja (lkt) *wv.* draw up, muster (an army) 17/19.
fylla (d) *wv.* fill 60/7 ('lets blood fill swords'?—see 60/8 n.); complete 10/9 (cf. **fullkominn**); expand, fill out, amplify 6/20.
fyllr *f.* that which fills, filling; *hjálms f.* = head, in kenning for sword: *Vindhlés hjálms f.* 7/1 (instr.) (cf. *Gylfaginning*, ch. 27, *Skáldskaparmál*, ch. 8); food 53/7.
fyr *prep.* with acc. (or dat.) before 52/5; of time, *fyr liðit ár* in the far past 94/6.
fyrðar *m. pl.* men 1/3 (dat. pl. with *bannat*), 25/2 (with *gefr*), 65/7 (with *eyddi*).
fyrir *prep.* (1) with dat. in front of 19/7, 64/4; *fyrir grundu* on the landscape 58/5; before, in s-one's presence 45/2, 68/6 (with *hers gnótt*), 80/2 (with *fjǫrum*), 80/8; because of, by the agency of (or in the presence of, with?) 41/3, 90/8; *vera fyrir e-u* have precedence over, be the principal kind of s-thing 1/43. (2) with acc., before, in front of, in the presence of 3/3, 36/10, 97/3; in the face of 24/7; past 35/6; because of, as a consequence of (in exchange for?) 45/3. (3) as adv., in front 54/11.
fyrirboðning *f.* prohibition 0/7.
fyrr *adv. comp.* before, earlier; in the past 51/10; above 2/10, 8/48, 17/21; *fyrr . . . en* sooner than 102/5.

fyrri *a.* earlier, first (of two) 7/9, 8/39, 15/9, 28/11, 39/9, 58/10; preceding 15/10, 16/9, 10, 27/10, 69/9, 93/9 n.
fyrst *adv.* first 35/10 (i.e. he was the first to invent), 67/11 (in the beginning, i.e. of *Háttatal*).
fyrstr *a. sup.* first 24/9, 10, 27/12, 28/9, 13.
fýsa (t) *wv.* urge; impers. *skip fýsir e-s* the ship(s) is (are) eager for, hasten(s) towards s-thing 20/8.
fættir *m.* diminisher, one who reduces the number of certain things (by giving them away); in kenning for generous ruler (Skúli), *sá menja f.* 45/4.
fœðir *m.* feeder, in kenning for war-leader: *f. gunnstara* 92/6 (obj. of *finnrat*).
fœra (ð) *wv.* bring, transfer 17/27; inf. (parallel to *stœra*) with *skal* (impers.), present (with *stef* as obj.) 81/8.
fǫr *f.* travelling, journey, expedition 38/1, 80/5 (dat. obj. of *ypt*), 92/1 (obj. of *geta*); *at fǫr* on the journey, in the travelling 44/16.
fǫt *n. pl.* clothing; in kenning for mail-coat(s) 53/6 (obj. of *slítr*).

gagn *n.* win(ning), victory 9/3, 59/8.
gagnstaðligr *a.* antithetical, of contrary or opposed meaning 16/14, 23/9 (cf. *TGT* 113, 114).
galdr (drs) *m.* chant; in kenning for battle, *skjaldar g.* 58/8.
galdralag *n.* incantation metre 100/7.
galli *m.* flaw, cause of damage, destruction; in kenning for wine, *strúgs g.*, that which robs a man of dignity (or anger?) 25/8 (subj. of *kemr*); for winter (cf. *Skáldskaparmál*, ch. 29), *orms g.* 83/5 (acc. of time).
gamall *a.* old; of mead 24/6 (with *ǫldur*).
ganga (1) *f.* travelling, movement (*við*: over) 23/4.
ganga (2) (gekk) *sv.* go, walk 64/3, 71/8; run, continue 81/9; *g. fyrir* stand for, be equivalent of 17/21.
garðr *m.* enclosure, fence; in kenning for shield-wall: *Þundar grindar jaðra g.* Þundr's gate's edges' fence 58/5.
gát *n.* food, prey 11/3.
gata (1) *f.* path 6/10.
gata (2) = *gat* + neg. suffix 90/7, see **geta** sv.
gátt *f.* door-opening 89/2.
geðveggr *m.* mind-wall, i.e. breast; in kenning for wound, *gluggi geð-veggjar* 50/5.
geimi *m.* ocean 13/8; *geima slóð* 71/2.
geirr *m.* spear; in kenning for battle, *duna geira* 53/2.
geisa (að) *wv.* rush 21/2 (with *frá*), 22/8, 52/5.
gella, gjalla (gall) *sv.* resound, ring 9/7, 60/4.
gelmir *m.* (= *gemlir*) hawk 2/7.

gengi *n.* accompaniment, company 26/8.
gera (ð) *wv.* make, cause to be, create 1/9, 8/19; form 28/11, 58/10; compose 92/2; achieve 40/3; constitute 1/25; do 8/46; give (details) 100/2; impers. with dat. *e-m gerir kátt* s-one becomes cheerful 89/1; pp. *gǫrr* performed 44/16.
geta (gat) *sv.* get, be gifted with 90/7 (with suffixed neg.); can find 84/6 (subj. *hinn er mál metr*; if this clause is taken with *milding*, *getr* must be impers., there can be found) 84/6; with gen. mention, speak of 70/5; impers. *getit var e-s* mention was made of s-thing 92/1; with inf. be able 37/7 (with *bjóða*), 101/5, with *at* and inf. 13/5 (1st pers. sg.), with pp. 51/3.
gilda (d) *wv.* make strong, increase, encourage; satisfy, do justice to? (cf. *NN* 1317); *g. gráð vargi*, i.e. by feeding it every day, by fighting battles 56/3. Cf. **auka, herða, hressa, remma, stœra**.
gildi *n.* feast 11/2 (to prepare a feast for wolves is to wage war).
gipt *f.* good luck, fortune (what is given by fate) 1/7.
gjald *n.* payment 37/2 (obj. of *fest*; according to *NN* 1311, parallel to *vald* and object of *banna*, with the meaning 'repayment, vengeance').
gjaldseiðr *m.* 'payment-coalfish', in kenning for *ægishjálmr*; *grundar gjaldseiðr* = serpent (fish of money of ground or fish of the ground where money is buried; fish whose ground is gold?) 15/2 n.
gjarn *a.* eager; comp. 55/3 (with *at*, to do s-thing).
gjǫf *f.* gift 26/6, 44/2 (collective); *at g.* as (for) a gift 23/7, 49/2, for giving, to be given 28/6 (with dat., to s-one), 86/6.
gjǫflati *m.* one who is slow to give (*e-s* s-thing), a miserly person (*e-s* with s-thing) 97/6 (gen. with *hróðr*).
gjǫflund *f.* disposition to give, generous nature 90/7 (obj. of *gata*).
glaðdript *f.* joyful snow, in kenning for gold 43/4.
glaðr *a.* happy 4/17, 17/8 (with *fylkir* or *ræsir*), 24, 86/5 (predicative, see **drekka**), 91/1 (with *ferð*); n. as adv. *er glatt* it is a joy, there is gladness 83/5; as subst. m. steed (from the name of a mythical horse (*Gylfaginning*, ch. 15, *Skáldskaparmál*, ch. 58), in kenning for ship: *g. Geitis* 44/15.
glamm *n.* noise, uproar 20/4.
gleðja (gladda) *wv.* gladden; *g. e-n e-u* i.e. make presents of s-thing to s-one 24/1, 46/3.
glóa (að) *wv.* shine, glow 72/2.
glóð *f.* red-hot embers; in kennings for gold (rings), *síks g.* 17/1, 9, 10, 27, *lýslóðar g.* 45/8; in kennings for sword, *Hlakkar g.* 50/3, *rimmu g.* 57/7, *styrjar g.* 85/7.
gluggi *m.* window; in kenning for wound, *geðveggjar g.* 50/5 (obj. of *svífr*).
glygg *n.* storm, strong wind 52/4 (subj. of *hrindr*), 77/7 (subj. of *náir*); in kenning for battle, *Gǫndlar g.* 59/7.
glymr *m.* noise, clash, in kenning for battle: *g. skjalda* 55/4.

glær (1) *a.* transparent, that can be seen through ? 50/6 (with *geðveggjar*; see next).
glær (2) *m.* sea; in kenning for blood, *g. hræs* ? 50/6 (part of kenning for sword; cf. **glær (1)**).
glæsa (t) *wv.* adorn 73/1.
gløggr *a.* precise, close 100/1.
gnapa (ð) *wv.* tower 83/1.
gnapturn *m.* towering (jutting) tower; in kenning for head, *g. aldrs* 50/8.
gnótt *f.* sufficiency, abundance, multitude 68/8 (with *fyrir*); great extent 93/8.
góðr *a.* (n. **gott**) good 51/8, 84/1 (with *orð*), 86/3.
góla *adv.* readily 23/7.
gómr *m.* gum (of mouth), in kenning for tongue 85/3.
gotnar *m. pl.* men 11/5, 91/2; dat. with *vensk* 44/1, acc. with *þrotna* or gen. with *elli* 61/8.
gráðr *m.* greed, hunger 56/4.
gramr (1) *a.* angry 18/12.
gramr (2) *m.* king, ruler ('hostile one') 1/7, 4/4, 18/6, 18, etc.; *of gram* 19/3, 22/6, 61/2; often refers to Skúli, 36/4, 37/8, 68/3 (dat. of advantage), etc.; dat. with *skotnar* 44/1, with *þakka* 46/7, gen. with *skrautfara* 70/5, with *spjǫll* 80/4; context illegible at 94/1.
grand *n.* injury, damage 57/8 (obj. of *remma*).
granda (að) *wv.* with dat., harm 18/8, 13, 20 (subj. is *brandr*, obj. *seggjum*).
grár *a.* grey (of wolf) 51/6.
greiða (dd) *wv.* smooth out, unfold, straighten 77/7.
grein *f.* distinction ('contrasting feature') 0/15 (variant *málsgrein*), 0/23, 1/26, 6/17, 18, 97/9; division, variant, type, category 4/21, 8/24 (i.e. *leyfi*), 27/14, 67/11; detail, account, particular 100/1. Cf. *TGT* 49, 101. The contrast in *tala ok grein* at 0/15 may be equivalent to 'quantitative and qualitative'.
greina (d) *wv.* distinguish 0/28, 1/36, 2/11; make a distinction in (between) 8/46, 31/11; analyse 1/28; divide 51/9, 58/17; diverge? signify? 16/12 (*at greina*: in their distinction?). Cf. *TGT* 49–50, 101.
greip *f.* grip, grasp of the hand 77/8, 91/5.
greizla *f.* payment, handing over; *er búinn til greizlu* i.e. is offered readily 88/5.
greppaminni *n.* 'poets' reminder' 39/12. Cf. *FoGT* 147.
greppr *m.* man 72/1 (subj. of *róa*); poet, = Snorri 70/6.
greypr *a.* rough, dangerous, cruel 71/4 (with *slóð*).
grið *n. pl.* peace, truth 9/6.
gríma *f.* helmet; *g. grundar gjaldseiðs* = *ægishjálmr*, (metaphorical) helmet of terror (cf. *Skáldskaparmál*, ch. 40) 15/1 (see *LP* s.v. *ægishjalmr*).

grimmr *a.* grim 35/6 (with *hrannláð*).
grimmsettr *a.* (*pp.*) grimly equipped (with *il*; referring to the claws on a wolf's paw) 64/8.
grind *f.* gate, in kenning for shield, *Þundar g.* 58/6.
grund *f.* ground; *fyrir grundu* on the landscape 58/5; in kennings for ruler = Norway 52/3 (gen. with *tiggja*), 90/8; for shield (in kenning for battle), *Hrundar g.* 61/5, (in kenning for warrior) *Mistar lauka g.* 85/4; for head, *heila g.* 63/8 (gen. with *meginundir* or gen. of respect); for serpent, *grundar gjaldseiðr* 15/1.
grunn *n.* shallow; as first half of compound separated by tmesis *grunnfall* shallow breaking water, shallows 19/4 (though *grunn* could be taken as a. n. pl. either with *fǫll*, or with *skip* if *djúp* is taken with *fǫll*; see *NN* 1303).
grunnr *a.* shallow 19/4 (?—see **grunn**), 19/11.
grýttr *a.* (*pp.*) covered with (surrounded by) boulders 15/5 (with *setri*).
grœðir *m.* gainer, increaser, gatherer (one who makes s-thing grow), with gen. in kenning for Skúli 81/2. Cf. *stœrir*.
grœnlenzkr *a.* of Greenland 70/17, 72/10.
grœnn *a.* green 30/2 (with *skjalda*).
grǫn *f.* lips, chops 11/8, 96/2.
guð *m.* God 12/2.
gull *n.* gold 2/7, 17/27, 23/8 (descriptive gen. with *skál*), 47/7 (dat. of respect with *viðr*), 89/8 (obj. of *brýtr*).
gullbroti *m.* gold-breaker, distributor of gold, generous ruler = Skúli 47/7, 88/6.
gullhættr *a.* dangerous to gold (by breaking it up and giving it away, being no lover of it), liberal with gold 47/8 (with *gullbroti*).
gullinn *a.* golden 91/5.
gullsendir *m.* gold-distributor, = Skúli 61/5.
gulr *a.* yellow (of shields) 8/6.
gumi *m.* man; gen. pl. 8/5, 83/6; dat. pl. *gumnum* 91/6 (with *sendir*).
gunnfani *m.* battle-flag 52/3 (obj. of *hrindr*).
gunnhættir *m.* battle-darer, warrior; = Hákon 15/5, = Skúli 43/3.
gunnseiðr *m.* 'battle-(coal)fish', kenning for sword 2/8.
gunnsnarr *a.* battle-keen 11/4.
gunnstari *m.* battle-starling, kenning for raven or eagle; in kenning for warrior, *fœðir gunnstara* 92/6.
gunntamr *a.* accustomed to battle, good at fighting 84/2.
gunnveggr *m.* battle-wall, kenning for shield; in kenning for warriors 61/6.
gunnǫrr *a.* battle-keen, -quick, or -generous 80/2 (with *fjǫrum*).
gylðir *m.* 'howler', wolf 11/3.
gæta (tt) *wv.* take care; pp. (sc. *hefir verit*?—or perhaps read *gæt* and take it as imp.) 42/9.

gætir *m.* guardian, protector 13/2 (dat. with *fá*).
gœða (dd) *wv.* enrich, add strength to, increase 55/3.
gǫfugr *a.* noble 3/6, 4/6, 16, 39/8 (with *ógnsvellir*).
gǫrr *pp.* of **gera**.
gørr *a.* (= *gǫrr*) with gen. ready for, full of 80/6 textual n. (with *fǫrum*).

hádýr *n.* high animal or thole-animal, in kenning for ship, *hranna hádýr* 8/35 (*hár* m. oar-thole, rowlock; *hádýr* would be a complete kenning for ship in itself).
haf *n.* sea 17/13 (subj. of *er*), 20/6, 93/5; instr. with *slétt* 17/2; with def. art. *hafit* 20/2, by word-play also pp. of *hefja*.
hafa (ð) *wv.* have 70/15; use 47/9; *h. til* use as, make serve as 45/10, 58/9; with pp. 14/2 (*hefir*), 41/3 (*hefr*), 67/5 (subj. is *jǫfrar*), 1st pers. sg. *hefi ek* 92/2, 100/2.
hafbekkr *m.* ocean-bench, kenning for ship, in kenning for sea 75/2.
hafhreinn *m.* sea-reindeer, kenning for ship 19/5 (indirect obj. of *viðr*).
haflauðr *n.* sea-foam 76/2 (acc.).
hafrǫst *f.* sea-current 74/1.
hagbál *n.* skilfully-made pyre, in kenning for (objects made of) gold: *lagar h.* 44/6 (gen. with *herfjǫlð*).
hagl *n.* hail, in kenning for (hail of) battle, missiles: *sóknar h.* 62/2 (obj. of *snýr*).
hagmæltr *a.* 'skilfully spoken'; n. as subst. as name of a verse-form 69/12, 71/9.
halda (helt) *sv.* with dat., hold, keep, defend 1/51 (inf. after *skal* 1/48), 3/6, 8/47, 53/3; inf. with *bið* 30/2; maintain 80/9 (pp.); be maintained 6/14, 48/9; hold, i.e. direct (subj. *vísi*, obj. *boga driptum*; alternatively it may be taken as impers. with dat., *e-u heldr* s-thing comes, drifts, with *regni*) 62/5 n.; word-play with **heldr** adv. 18/9; impers. pass. *e-u er haldit* s-thing is maintained 43/9 (sc. *minni alhendu*), 47/10; *h. aptr* hold (s-thing) back 17/23; *h. til* be in accordance with 6/10; md. for pass. 54/2 ('cannot be held'; or it may be reflexive, 'they cannot defend themselves').
hálfhneptr *a.* 'half-curtailed' 92/10; n. as name of a verse-form 76/12.
halr *m.* man 66/5.
hamdøkkr *a.* dark-coated (of raven) 5/5.
hams *m.* slough of snake 6/11.
happ *n.* good fortune, success 40/3, 7, 93/1; dat. with *næstir* 99/8.
hár *a.* high 22/1; deep (or with high waves?) 35/3; noble, great 4/4 (paraphrased *mikill* 4/14), 27/6, 61/2 (with *tími*); comp. *hæra* 98/5 (with *hróðrar*). Cf. 37/5–8 n.
háraustr *a.* high-voiced, noisy 19/6 (with *Rán*).
harða *adv.* mightily 4/5, 20/3 (with *svífr*); very 13/7 (with *hæf*; unless it is the a. with *ferð*); violently 57/6.

harðliga *adv.* mightily 4/15.
harðr *a.* hard 5/10, (?)13/7, 18/4 (with *skjǫldr*), 18/16; firm 18/11; of syllables, strong? accented? 0/29 n. (cf. **linr**).
harðráðr *a.* firm-ruling 12/2.
harðsveipaðr *a.* (*pp.*) tightly twisted or drawn, firmly rigged 34/6 (sup.).
harri *m.* lord; = Hákon 11/4, 23/1; = Skúli 53/5, 68/7 (dat. with *kærr*).
hástallr *m.* high stand; *hlunns h.* high slipway-stand, or high supporting structure (for a beached ship) 83/3.
hata (að) *wv.* with dat., hate, be opposed to 17/10, 19/11; be dangerous to, strive with 19/4; be hostile to, i.e. give away, be generous with (or referring to ruler breaking gold?) 90/6.
hati *m.* hater, enemy, in kenning for generous man who does not love gold but breaks it and gives it away: *ægis báls h.* 3/2.
háttafǫll *n. pl.* metrical fault or inconsistency (= Latin *lapsus metri*?) 51/11, 53/12, 58/15 (used of variation of metre within a stanza; cf. *FoGT* 147).
háttaskipti *n.* change of verse-form 27/13, 33/9.
háttatal *n.* number or list of verse-forms or stanza-types 11/10.
háttlausa *f.* lack of form (i.e. there are no hendings, the verse is *hendinga-lauss*) 54/9, 55/9, 56/9, 66/9, 10.
háttr *m.* (dat. *hætti*, pl. *hættir*, acc. *háttu*) manner, mode; manner of behaviour, how to behave 37/5 (obj. of *bjóða*; cf. 37/5–8 n.); characteristic behaviour 89/5; fashion, manner (or metre?) 95/6 (sg.); *at hætti* as usual (or 'in accordance with the metre'?) 37/10; category, types of usage 2/11; verse-form 1/9, 25, 6/18 etc.; pl. 0/4, 29/9, 10, 30/9; gen. pl. after num. 67/4, acc. pl. 31/11, 100/6; variation of form 8/15; *í (af) fyrra hætti* in (from) the (or a) preceding verse-form 69/9, 93/9 n. See note to 0/1 and cf. *FoGT* 136, 137, 138, 147, 150.
haukr *m.* hawk; in kenning for raven, *Hlakkar h.* 5/6 (dat. of respect).
hauss *m.* skull 10/4, 51/4, 65/2.
heðan frá *adv.* from now on 27/13.
hefja (hóf) *sv.* lift; enhance, raise to glory? 39/1; impers. *e-t hefr upp* s-thing begins 14/9, 27/9, 30/11; md., begin 19/14; pp. *hafit* (word-play with **haf** n.) 20/2.
hegna (gnd) *wv.* punish 17/30; keep back, hold in check 24/3.
heiðfrǫmuðr *m.* payment-performer, one who honours (s-one) with payment of salary, generous lord, = Skúli 48/7.
heiðgjǫf *f.* gift of payment, salary 48/8 (dat. with *venr*).
heiðir *m.* hawk; in kenning for arms, *heiðis reiðir* 48/5.
heiðmaðr *m.* a man in the pay of a king or earl, retainer, soldier 48/5 (dat. of respect or advantage with *býr reiðir heiðis*).
heiðmildr *a.* generous with payment (of salary) 48/6.

heiðr *a.* bright, splendid 14/1 (with *orðróm*), 28/4 (with *gjǫf*), 48/7 (with *leiðar*: they are bright because of their gold ornaments).

heili *m.* brain, in kennings for head: *heila bœr* 2/6, *heila grund* 63/8.

heill *a.* whole, complete 13/12 (cf. **fullkominn** and *FoGT* 137); entire, unbroken 19/8 (with *hrǫnn*), 19/11.

heilsa *f.* salvation, in a kenning for mead (which aids loquacity): *h. máls* 25/5.

heim *adv.* (at) home; with *sœkja* 13/8, 93/6.

heimdregi *m.* stay-at-home 98/6 (parallel to 98/3–4).

heimr *m.* world; pl. 96/8.

heimsvist *f.* staying at s-one's home, visit (with dat. of person visited) 29/2.

heit *n.* promise or threat 17/4, 30 (see 17/3 n.).

heitfastr *a.* true to one's word 35/3 (with *hjálm-Týr*).

heiti *n.* (descriptive) name, name which is not the normal one by which an object or person is called 2/9; with gen. 12/4, 11, 18/13, 27/5.

heitr *a.* hot 17/16.

héla (d) *wv.* freeze 20/1.

heldr *adv.* rather 18/2, 9 (word-play with pres. of *halda*); quite, i.e. extensively 8/3 (with *lituð*).

helmingr *m.* half-stanza (four lines) = *vísuhelmingr* 8/38, 19/10, 22/9, etc. (cf. *FoGT* 136–7).

henda (nd) *wv.* catch 17/16 (word-play with *hendir* 'thrower').

hending *f.* rhyme, assonance (both internal rhyme and end-rhyme) 0/30, 1/37, 2/9, etc.; rhyme-syllable 38/10, 49/10, rhyme-word 76/11, 77/12, 79/13; *at hendingum* as regards internal rhyme 23/12; *standa í hendingum* form part of the rhyme-scheme 44/12; *með einni h.* with the same rhyme 80/9; *sér um h.* with a separate rhyme 82/10. In *SGT* the word refers to the combination or conjunction of particular vowels and consonants.

hendingalauss *a.* without hendings, rhymeless 69/10, 72/9; n. as adv. 76/9.

hendingarorð *n.* rhyme-word 58/11.

hendingaskipti *n.* change or distribution of rhymes 27/15 (cf. **stafaskipti**).

hendir *m.* thrower, in kenning for generous ruler, *h. loga ǫldu* distributor of gold 17/3, 15 (word-play with pres. of *henda*), 17/29.

heppinn *a.* fortunate 38/5.

hér *adv.* here; i.e. in this verse 8/20.

herða (ð) *wv.* make hard, give strength to, increase 57/2; impers. *e-t herðir* s-thing hardens, becomes taut 21/3 (inf. with *frá*).

herðar *f. pl.* shoulders; in kenning for head 65/4.

herðir *m.* hardener, increaser, promoter; in kenning for ruler (Skúli), *fólka h.* 37/3 (dat. with *fest*).

herfang *n.* war-gear; pl., obj. of *hristi* 36/2.

herfjǫlð *f.* great number, quantity (with gen.); obj. of *bar* 35/7 (with *brims blakka*); great heap, subj. of *friðask* (or acc. with *við* if *valdi* is taken as subj.) 44/5 (with *hagbáls*).

herframr *a.* war-bold 75/5 textual n.

hergramr *a.* war-fierce 75/5 (with *rógálfr*).

herkaldr *a.* extremely cold 61/4.

herr *m.* army 18/5, 18, 21/3, 49/4, 56/7, 68/8 (gen. with *gnótt*), 88/2; dat. with *reyndisk* 36/5; dat. of respect with *er væni* 58/2 ('the armies can expect'); men 87/5 (dat. of respect with *sker*).

hersir *m.* lord (i.e. of a *herað*, 'district', or of a *herr*?), a high rank in Norway 7/7 (dat. of respect), 29/7 (gen. with *stillir*); as a title, = lord 27/5.

herstefnir *m.* war-leader, = Skúli 51/1.

hertogi *m.* war-leader, = Skúli 40/5, 66/2; as title, duke 0/3. (Probably not used as a title in the verses. Skúli only acquired that title in 1237, long after *Háttatal* was composed. But the heading to *Háttatal* in U was probably added by a scribe after that date.)

hestr *m.* horse; in kennings for ship, *h. rasta* 34/5, *h. lesta* 44/17, *h. festa* 71/6, *h. svanfjalla* 83/4 (subject of *gnapa*).

heyjask (háðisk) *wv. md.* be held, be waged 18/1, 16.

heyra (ð) *wv.* hear 97/5 (with acc. and pp.).

hildigǫltr *m.* helmet ('battle-boar'; cf. *Skáldskaparmál*, ch. 44); dat. sg. (instr.) *-gelti* 2/5. The word may relate to the 'boar-images' used on early medieval helmets and referred to in Old English poetry, and presumably means 'helmet' by synecdoche; unless the reference is literally to protecting the head with a boar-shaped piece on the helmet. See *NN* 1295.

hildr *f.* battle 10/6, 40/1, 53/3.

hilmir *m.* ruler ('helmet-provider' or 'leader of helmeted men'?); = Hákon 7/1, 5, 8/1, 29/1 (dat. with *heimsvistir*), 30/1; = Ingi Bárðarson 33/5 (*hilmis bróðir* = Skúli), 39/5 (with *foldar*); = Skúli 34/5, 52/8, 62/7, 95/2 (with *Mæra*; dat. with *flutak*).

himinn *m.* heaven, sky 95/8.

hinn er *pron.* (he) who 8/34 (with *stýrir*), 24/7 (with *skjǫldungr*), 71/5 (with *leyfðr skati*), 84/5 (with *milding*; or 'he who' as subj. of *getr*).

hirð *f.* court, ruler's following; obj. of *laða* 86/4, of *kallar* 88/1, dat. obj. of *gerir* 89/1; dat. with *hollastir* 99/7.

hirðmaðr *m.* member of the *hirð* 75/6 (obj. of *lítr*).

hitna (að) *wv.* become hot 9/4.

hitta (tt) *wv.* meet, come across; impers. *eigi hittir* there cannot be found 91/7.

hjaldr (ldrs) *m.* noise; battle 39/5 (gen. with *stýri*), 40/5, 94/3 (obj. of *efla*); in kenning for sword, *hjaldrs hyrr* 50/8.

hjaldreifr *a.* battle-happy 39/1.

hjaldrremmir *m.* battle-strengthener, = Skúli 49/1.

hjaldr-Týr *m.* battle-Týr, kenning for war-leader (Skúli) 53/4.

hjálmlestandi, hjálmlestir *m.* helmet-damager, warrior 49/6 (dat. with *býr*), 59/1 (= Skúli).
hjálmr *m.* helmet 8/1, 39/4 (pl., subject of *hófu*); in kenning for head 7/1, for battle 57/1, for sword, *hjálma hyrr* 58/1.
hjálmsvell *n.* helmet-floe (-ice), kenning for sword 60/4 (obj. of *lætr*).
hjálm-Týr *m.* helmet-Týr, kenning for warrior (Skúli) 35/4.
hjarni *m.* brain; in kenning for head 64/8.
hjarta *n.* heart 4/8 (dat. with *una*), 4/18; i.e. valour 15/7.
hjástæltr *a.* (*pp.*) 'abutted'; n. as subst. as name of a verse-form 13/1 textual n., 13/9.
hjǫrr *m.* sword 7/3 (instr.), 8/2, 10/8 (instr.), 36/6 (subject of *lék*), 62/4; in kennings for warrior(s), *hjǫrs rjóðr* 41/5, *hjarar Baldr* 43/2, *hjarar draugar* 47/6, *hjarar lundar* 60/8.
hjǫrtr *m.* hart, stag; in kenning for ships, *stinga h.* 73/8 (dat. of respect).
hlaða (hlóð) *sv.* load 76/1 (pp.).
hlaupa (hlaup) *sv.* run, gallop; of a ship (or ships) described as Róði's horse (stud) 21/4.
hlekkr *m.* link in a chain; fetter; *handa h.* = arm-ring 87/1 (collective); *Manar h.* (that which surrounds M.) = sea 77/2.
hlemmidrífa *f.* noisy snowstorm; *h. Hildar* = rain of weapons 54/3 (dat. obj. of *skýtr*).
hleypa (t) *wv.* cause to gallop (with dat. of the animal, acc. of the ground galloped over) 71/4, 76/5 (object is *skíðum*, i.e. ships). For the construction cf. **renna (2)**.
hlið *f.* side 38/8.
hlíð *f.* slope, hill(side); *fjǫrnis h.* = head 65/6 (acc. with *rendi*, over; collective).
hlíf *f.* shield 4/3, 13; *hregg hlífa* = battle 62/3.
hlífgrandi *m.* shield- or defence-damager, i.e. sword 17/3 (instr. with *rekr*), 17/14, 29.
hljóð *n.* sound 0/28 (obj. of *greinir*), 1/25, 47; gen. pl. (with *-setning*, parallel to *stafa-*, referring to hendings) 1/27; 27/14 (referring to alliteration?); *hljóðin ǫll* (referring to alliteration and hendings) 65/12.
hljóðfall *n.* consonancy, assonance 1/32.
hljóðfyllandi *m.* alliterating syllable (in odd lines) 28/10, 69/10, 85/10.
hljóðfylling *f.* = *hljóðfyllandi* (cf. **stuðill**) 39/10, 68/13.
hljóðsgrein *f.* distinction of sound 0/26, 29–30. Cf. *SGT* 50, 52; *TGT* 33, 49, 50, 54–5, 63, 66–7 (= *accentus*, *tenor*, *tonus*).
hljóðstafr *m.* vowel 1/20 (complement), 21, 22, 31, 32, 35, 8/20; the meaning is uncertain at 32/10, see n.; in *h. sá er kveðandi ræðr* (82/11, see n.) it seems to refer to the alliteration, though neither *hljóðstafr* nor *ljóðstafr* seems to be used elsewhere in medieval texts for 'alliterating stave' (in the text of *TGT* 54 in W the word is probably a scribal error);

cf. Konráð Gíslason (1875). In *FGT*, *SGT*, *TGT*, *hljóðstafr* means 'vowel'.

hljóta (hlaut) *sv*. get (as one's lot), be fortunate enough to (or 'have to'?), with inf. 70/5, with *at* and inf. 29/3; be forced to (with inf.) 33/8, 34/7.

hlumr (or **hlummr**) *m*. oar-handle 23/2; pl., obj. of *spenna* 75/5 (normalizing to *hlummi* would make the metre of this line conform to that of 75/1).

hlunnr *m*. roller(s) (part of slipway for ships), planks forming slipway 71/5; perhaps the props for beached ships, stocks 83/3; in kenning for ships, *hlunna skíð* 76/7.

hlunnvigg *n*. slipway-steed, i.e. ship 74/2.

hluthenda, hluthending *f*. mid-rhyme, a rhyme not at the beginning of a line 1/41, 8/13, 63/9.

hlýða (dd) *wv*. be permitted, acceptable 1/22, 8/12.

hlymr *m*. din; in kenning for battle, *h. Gungnis* 52/6.

hlynr *m*. maple (tree); (*Huglar*) *rasta h.* = ship 19/1, *Hristar h.* = warrior (Skúli; phrase belongs with *fellr*) 61/3.

hlýr *n*. bow (of ship) 20/1 (by word-play also 3rd pers. sg. of *hlýja* wv. to make warm); subj. of *náir* 76/4.

hlǫkk *f*. battle (also the name of a valkyrie) 18/1, 15; in kenning for warrior, war-leader, *hlakkar snerpir* 42/2, = Skúli. Cf. *Hlǫkk* in Index.

hneptr *a*. (pp. of *hneppa* wv.) 'squeezed', reduced, i.e. shortened by one syllable, monosyllabic 77/12, 78/10, 81/10, 84/9, 87/9, 89/9, 91/9, 93/9. Cf. **stýfa**.

hníga (hneig) *sv*. sink down; *aldir h. í gras* 16/2; *h. und e-t* bend down (under s-thing), move (underneath s-thing) 49/3.

hnigfákr *m*. bending, bucking horse; *Haka h.* = ship 71/3 (dat. obj. of *hleypa*). (Fákr is the name of a mythical horse in *Skáldskaparmál*, ch. 58; its rider happens to be Haki!)

hniggrund *f*. dipping, tipping ground; *hafbekks h.* = sea 75/1 (obj. of *hrinda*).

hnigþili *n*. bending board, pliable plank; *h. randa* = shield(s) (acc. with *brjóta*) 59/4 ('rim-boards' could be a kenning for shields, or could mean the actual boards of which the shields are made).

hnoss *f*. jewel, treasure 29/5; gen. with *gjǫflata* 97/8.

hnugghendr *a*. 'deprived-rhymed'; n. as subst. 75/13.

hnykkja (kkt) *wv*. with dat., wrench (off) 57/5.

hoddgrimmr *a*. hateful, cruel, to hoards, i.e. generous with gifts 66/6.

hoddspennir *m*. hoard-spender, i.e. generous man (or 'hoard-acquirer, -grasper'?), = Snorri 29/8.

hoddstiklandi *m*. hoard-flinger, i.e. generous ruler (= Skúli) 39/2. Cf. **stiklir**.

hof *n*. temple; in kenning for breast, *hugtúns h.* 50/3.

hold *n*. flesh 65/8 (dat. with *fenta*).

hollr *a.* gracious 18/6, 11, 18, with dat., loyal, devoted (to s-one) 29/7; sup. with dat. 99/7.

holt *n.* bare hill (or wood); in kenning for (top of) head (or for the hair? Cf. *NN* 1295), *h. heila bœs* 2/5.

horn *n.* (drinking) horn 24/2; in kennings for beer or mead 24/1, 25/4.

hraða *adv.* quickly 43/4.

hraði *m.* speeder, flinger; *hrings h.* = generous prince (Skúli) 86/1 (acc. with *veit*).

hraðsveldr *a.* (*pp.*) quickly-swollen, swift-growing 55/4.

hrafn *m.* raven 51/1; dat. of respect (advantage) 32/6.

hrannláð *n.* wave-land, i.e. sea 35/8 (subject of *bar*).

hraustr *a.* valiant, vigorous 62/7.

hregg *n.* storm; in kennings for battle, *Hrundar grundar h.* 61/6 (dat. for locative), *h. hlífa* 62/3.

hreggǫld *f.* storm-time; *Hristar h.* i.e. time of battle 59/2 (dat. of time, i.e. in battle).

hreingullinn *a.* pure gold 87/6.

hreinn (1) *a.* clean, polished, shining 4/5, 14 (complement), 8/7 (with *þrǫmu*; cf. 8/8 n.).

hreinn (2) *m.* reindeer; in kenning for ship, *Røkkva reina h.* 73/4 (obj. of *glæsa*).

hreintjǫrn *f.* pure lake or pool; in kenning for drinks of ale or mead, *h. horna* 24/1 (instr.).

hressa (st) *wv.* make flourish, strengthen; *lætr hǫpp hrest* causes successes to be increased, i.e. to take place 93/1. Cf. **gilda**.

hreyta (tt) *wv.* fling, scatter; impers. pass. with dat., *e-u er hreytt* s-thing is scattered, flung (of generous giving) 41/7.

hreytir *m.* scatterer; *hringa h.* = generous prince (Skúli) 73/5.

hríð *f.* storm; in kenning for battle, *vápna h.* 65/1 (subj. of *náði*).

hríðfeldr *a.* stormy 76/8 (with *fjǫll*, i.e. waves).

hrími *m.* rime, frost, dew; in kenning for battle (referring to missiles falling like dew?—cf. **frost** and see Meissner 1921, 182), *hræljóma h.* 61/1.

hrinda (hratt) *sv.* push 52/3, 75/1.

hringdropi *m.* ring-droplet = gold (ring) 42/4 (cf. *Gylfaginning*, ch. 49, *Skáldskaparmál*, chs. 32, 35).

hringmildr *a.* liberal with rings (gold), i.e. generous 47/3 (with *hringskemmi*).

hringr *m.* ring (of coat of mail) 2/3 (instr.); arm-ring, gold ring 45/2 (gen. pl. with *brot*), 47/3 (dat. obj. of *þinga*), 49/2, 99/3 (dat. with *hæztir*); in kennings for (generous) prince 63/6, 73/7, 86/1, 90/8.

hringserkr *m.* mail-shirt 36/4 (obj. of *samði*).

hringskemmir *m.* ring-damager (-shortener?), i.e. generous giver, prince, Skúli 47/4 (acc. with *spyr*).

hrista (st) *wv*. shake 74/1; impers. (*herfǫng* is the obj.), i.e. armour was shaken 36/1.

hrjóða (hrauð) *sv*. with dat. clear away, get rid of; *h. auð* i.e. give generously 37/6; strip (*e-m e-u* s-one of s-thing), take away (s-thing from s-one) 33/5.

hrjóta (hraut) *sv*. fly, be flung 57/3.

hróðr (rs or **rar)** *m*. praise, fame, glory 68/6 (obj. of *bera*), 80/6 (gen. with *ǫrum*), 98/5 (gen. with *verðari*), 100/4 (gen. with *ørverðr*); (poem of) praise 81/7 (obj. of *hrœra*), 97/6 (obj. of *heyra*).

hrunhenda = *runhenda* 86/11 textual n., 88/9 textual n.

hrynhenda *f*. falling-, flowing-, or clanging-rhyme form 62/16, 63/11, 64/9.

hrynhendr *a*. falling-, flowing-, or clanging-rhymed 62/13, 63/12, 90/9.

hrynja (hrunða) *wv*. fall; ring, clang 57/4.

hrynjandi (1) *pres. p.* (*a.*) (= *hrynhendr*) falling, flowing, or clanging 61/10. The metaphor may imply a comparison with the sound of waves breaking on the shore (cf. Hallvard Lie in *KLNM* vii. 28, 'Hrynhent').

hrynjandi (2) *f*. (= *hrynhenda*) falling, flowing, or clanging verse-form 62/15.

hræ *n*. corpse; in kennings for blood 7/4, 32/4, 50/6 (or for sword, see **glær**), 60/3 n.

hræljómi *m*. corpse-light, i.e. sword; in kenning for battle 61/1.

hrænaðr *m*. corpse-adder, i.e. spear 79/8 (gen. pl. with *oddum*).

hrœra (ð) *wv*. stir, move, set in motion, i.e. perform, create (of poetry) 31/8, 81/7.

hrǫnn *f*. wave 19/7, 22/2 (obj. of *skilja*), 38/7, 76/1, 78/1 (obj. of *skerr*); in kenning for ships 8/34; in kenning for blood, *hræs fella hrannir* 60/3 n.; in kenning for (the mead of) poetry, *Hárs saltunnu hrannir* 31/7.

húfr *m*. the side of a ship, the planks (collective) 19/7, 35/7, 78/2 (subj. of *skerr*), 93/5.

hugdýrr *a*. noble-minded 39/6 (with *stýri*).

hugr *m*. thought; in kenning for breast 6/13 (= *sefi* 6/8).

hugrakkr *a*. bold in heart; sup. 99/4.

hugtún *n*. thought-enclosure, thought-field; *hugtúns hof* = breast 50/4.

hunang *n*. honey 24/8.

hungr (rs) *m*. hunger; gen. of respect 51/2.

húnskript *f*. mast-head tapestry, i.e. sail 78/8 (dat. with *ósvipt*).

hvar *adv*. where 42/3 (as conj., = *þar er*? or with *veit*).

hvarmr *m*. eyelid; in kenning for tears, *fagrregn hvarma* 42/8.

hvass *a*. sharp 32/3, 33/6, 63/7; n. as adv. 4/3, 13, (i.e. fast) 19/2.

hvatr *a*. sharp, keen 8/8; bold 5/11, 7/2; n. as adv., violently 36/1, energetically, vigorously, without holding back 41/8; hard 60/3 textual n., fast 78/1.

hverr *a. pron*. each, every 53/5 (with *snerru*), 84/8; each person 26/8; *í* *ǫðru hverju* in every other (alternate) 18/9; interrog., who 40/1–4, 55/1, 56/1; which, what 97/5 (with *seggr*).

hvetja (hvatta) *wv.* urge 15/8 (abs.); make keen, sharpen 50/7 (pp. with *at* and inf.: in order to, so as to).

hvíla *f.* bed; metaphorical for death on the battlefield, the bed prepared by Hildr 49/5.

hvíld *f.* rest (*at*: after) 23/4.

hvítr *a.* white; of weapons, shining 18/5 (with *fal*), 18/18, 54/4 (with *þrǫm*), 57/3, 73/3 (with *rítum*).

hylja (hulða) *wv.* cover 2/4, 43/7.

hylli *f.* favour, grace 30/1 (obj. of *halda*).

hyrr *m.* fire; in kennings for sword, *hjaldrs h.* 50/7, *hjálma h.* 58/1 (gen. with *styrjar*).

hýrr *a.* cheerful, friendly 29/1.

hæfr *a.* fitting, suitable 13/7 (complement).

hættir *m.* darer, in kenning for warrior: *vígs h.*, = Skúli 41/7 (see **at (1)**).

hættr *a.* dangerous (*e-u* to s-thing); sup. *hæztir hringum*, i.e. most generous 99/3.

hœgindi *n.* comfort, convenience, expediency 16/15.

hǫfuð *n.* head; dat. *hǫfði skemra* shorter by a head, i.e. (to be) decapitated 66/5.

hǫfuðskáld *n.* chief poet, major poet, 'classical poet' 0/20.

hǫfuðstafr *m.* chief letter, chief alliterating stave 1/13, 15, 20, 54/10, 68/10, 13, 69/11, 85/9, 97/10, 11, 12.

hǫfugr *a.* heavy 4/10.

hǫfundr *m.* judge 37/2 (subject of *kann*).

hǫgg *n.* blow 65/3; *í hǫggum* amid the blows 32/3.

hǫlðr *m.* yeoman, free farmer, man, subject 12/1 (obj. of *veldr*, parallel to *heiti*), 10, 37/5 (gen. with *áttar*), 44/5 (subj. of *bera*).

hǫll *f.* hall 23/6 (also by word-play f. of *hallr* a., 'askew, crooked, aslant'), 86/2, 88/1, 89/2.

hǫnd *f.* hand, arm (usually sg. collective) 23/1 (obj. of *firrask*), 26/2, 36/5 (obj. of *lék*), 45/7, 87/1.

íð *f.* labour, task; *gǫr er íð at* they work hard at 44/16 ('their journey is pursued hard').

iðja *f.* activity, work 20/7 (subj. of *lýsa*).

iðn *f.* labour, activity 9/1.

iðurmæltr *a.* (*pp.*) repeatedly said; n. as subst. 46/12 (cf. *FoGT* 147).

il *f.* sole (of the foot) 64/8 (of a wolf; to put heads under the sole of the wolf's foot = kill so that wolves can eat); in kenning for shield 30/4.

illr *a.* bad; *hafa illt til* be badly off for 35/12.

ilsporna (að) *wv.* tread underfoot 51/3.

ilstafn *m*. sole-stem (of a bird), i.e. claw 32/6.
ískaldr *a*. ice-cold 27/1 (with *ǫldu*).
íss *m*. ice; in kenning for swords, *álmdrósar ísar* 60/1 (obj. of *skylr*).
ítr (rs) *a*. noble, splendid 3/2 (with *hati*; or as first part of a compound *ítrbáls*, cf. *hagbál*), 4/7 (paraphrased *kostigr* 4/17), 13/3, 27/8 (with *auðs*), 30/7 (with *elli*), 52/8, 66/2, 101/3.
ítrhugaðr *a*. splendid-thoughted, noble-minded 41/1.

jaðarr *m*. edge, rim (of shield) 58/6; coast 79/8 (*með*: along); protection: *fólks j*. = Skúli 55/7.
jafn *a*. even 17/12.
jafnlangr *a*. the same length, of equal length 70/15.
jafnmargr *a*. the same number (of), as many (*sem*: as) 88/9.
jara *f*. fray, battle; in kenning for warriors 53/8.
jarl *m*. earl, ruler next in rank to a king (generally refers to Skúli) 27/7 (gen. pl. with *prýði*), 31/5 (dat. with *óð*: for), 55/2 (obj. of *séi*), 77/3 (subject of *lætr*, or of *sér* if the subj. of *lætr* is understood from the preceding stanza), 82/5 (part. gen. with *dýrstr*; perhaps parallel with *skatna* and to be taken with with both superlatives), 91/8 (part. gen. with *beztr*), 93/8 (gen. with *frama*); as title 94/8.
jarldómr *m*. earldom 39/8 (obj. of *gaf*).
jarlmegin *n*. power of the (an) earl (sc. Skúli) 54/8.
járngrár *a*. iron-grey 7/8.
jǫfurr *m*. prince 26/4, 55/1; = Hákon 1/8 (in apposition with *stillir*), 6/2, 10/6; = Skúli 37/6, 41/2, 42/6, 48/6, 51/8, 60/4, 65/8 (dat. with *kent*), 66/6, 72/5, 93/6; pl. = Hákon and Skúli 67/8 (subj. of *hafa*), 98/7 (obj. of *veit*), 99/1 (subject of *ró*).
jǫrð *f*. earth, land 12/2 (obj. of *lér*), 13/8, 17/2 (gen. with *skarð*), 12, 13 (at the second occurrence is subj.), 14, 28; coast, shore 35/6, 63/3; = Norway 1/3, 10/1, 16/8, 79/7 (obj. of *þyrna*).
jǫstr *m*. yeast (in kenning for ale) 25/1.

kaldr *a*. cold 17/4 (with *heit*), 22/8 (with *uðr*), etc.
kall *n*. call, naming; *at kalli* by name, said to be (i.e. they count as) 19/9, 12.
kalla (að) *wv*. call 0/30, 1/12, with suffixed pron. 25/8 ('this is the name I give to wine'); summon (*til* to) 88/2; with predicative a., describe as 4/12 (subj. is *egg*), 4/14, 5/9; *k. menn* is known as 9/12, 31/12; *svá at k*. thus, to say that 18/14; *má kalla* can be called 83/8.
kasta (að) *wv*. with dat., throw (aside) 76/3.
kátr *a*. cheerful; n. *kátt* 89/1.
kenna (nd) *wv*. teach; *k. e-m* with inf. force s-one to do s-thing 63/1; to use kennings 1/53; *k. rétt* use an appropriate modification (determinant) 6/9;

pp. *kent e-m* attributed to, belonging to s-one (with *sverð*) 65/8; which use kennings 1/55, provided with determinants 2/9 (cf. *vkenndir*, *TGT* 103; *Skáldskaparmál*, ch. 54); indicated, denoted, expressed by a kenning 8/38.

kenning *f.* modification, determinant 8/29; periphrastic description 2/10, 11, 12, 6/19, 17/21.

ker *n.* goblet 87/6 (subject of *leikr*).

kimblaband *n.* 'bundle-bond' 58/17, 59/11, 60/9, 10, 61/9.

kjǫlr *m.* keel 22/7, 38/8; dat. sg. *kili* 101/6 (instr.), 101/7 (obj. of *renna*); in kenning for sea 76/3, for ship 77/6.

klettr *m.* cliff, crag; in kennings for head, *hjarna k.* 64/8 (dat. obj. of *skaut*), *herða k.* 65/4.

klífa (kleif) *sv.* climb (upon) 51/6.

klifat *pp.* repeated; n. as subst. 47/12.

kljúfa (klauf) *sv.* cleave 8/1 (pp.).

klofna (að) *wv.* intrans., break, be cloven 19/8 (subject is *hrǫnn*), 19/12.

klóloðinn *pp.* (*a.*) shaggy (prickly) with claws 56/8 (cf. Fidjestøl 1982, 72–3).

klæði *n. pl.* clothes, clothing 35/12.

kløkkr *a.* flexible, pliant, yielding 21/8 (apparently with *stál*, but equivalent to adv., softly), 22/7.

kná *pret.-pres. vb.* auxiliary with inf., can, does 7/3, 15/5 (with *stýra*), 39/3, 60/5 (with *venja*), 72/1 (with *glóa*); impers. *sjá kná* one can see, there can be seen 58/5; with *at* and inf. 6/5.

koma (kom) *sv.* come 25/7; pp. having escaped 35/11.

konungdómr *m.* kingdom 14/8 (obj. of *ræðr* 14/1).

konungr *m.* king; = Hákon 1/4 (subject of *kann*, or of *lætr* if the subj. of *kann* is understood), 101/2, 102/3; = Ingi Bárðarson 69/2; as title, = Ragnarr loðbrók 53/14.

kostigr *a.* splendid 4/17 (cf. **ítr**).

kuðr (kunnr) *a.* known (*e-m* to s-one) 69/1 (with *kvæði*); *k. at e-u* known for s-thing 27/4.

kunnusta *f.* ability, art, artistry 6/20.

kveða (kvað) *sv.* say (in verse), recite, compose (poetry) 0/30 (pp.), 6/14 (pp.), 8/20, 15/9 (pp.), 16/17 (pp.), 17/9 (pp.), 44/13, 97/7 (pp.); compose in (a metre) 53/11 (pp.); impers. *er svá kveðr at* which sounds thus (or when it is composed thus?) 1/23 (*kveða at* = 'pronounce' in *FGT* 228, 230; cf. *atkvæði* 'pronunciation' in *SGT* 52, 66); *k. eptir* compose a poem using the story of 35/13 (pp.).

kveðandi *f.* poetical effect, sound of poetry; referring to alliteration (cf. *TGT* 68, 69, 96, 97) 1/9, 13, 54/10; referring to hendings (cf. *TGT* 64, 91) 1/26; it is uncertain precisely what is referred to at 82/11 (see 82/10–12 n.); referring more generally to recitation, poetical effect 4/19, poetical form?

8/17. Cf. *Laxdœla saga*, ch. 37, *ÍF* v. 106, *Skáldskaparmál*, ch. 72. The word appears to mean 'metre' in *FGT* 226, *TGT* 65, 88, 89 and *FoGT* 136.

kveðskapr *m.* composition (of poetry) 0/19.

kviða *f.* poem 35/12 (see 35/11 n.).

kvæði *n.* poem, series of stanzas 30/11, 35/12, 44/10, 53/13, 67/10, 13, 14, 69/1 (pl.), 70/13, 14, 95/3, 102/4.

kyn *n.* kin, kind, race 11/3 (acc. with *lætr*).

kærr *a.* dear 68/7.

kœnn *a.* clever 60/3 (with *jǫfurr*).

lá *f.* sea (surf) 21/2 (also by word-play p. of *liggja* 'lie'); 78/3 (obj. of *brýtr*).

laða (að) *wv.* invite 86/2 (inf. with *veit*).

lag *n.* tune; stroke (in rowing); manner, arrangement, measure, metre, verse-form 70/2, 78/11, 96/9, 97/12, 13.

land *n.* land 1/8 (obj. of *ræðr*), 17/13 (complement); = Norway 64/5; pl. 14/3, 30/6 (gen. pl. with *njót*), 79/2 (gen. pl. with *útstrandir*); in kenning for head (collective): *svarðar l.* 57/6 (obj. of *hnykkja*), for shield (in kenning for spear or sword): *fleina l.* 65/5 (with *meginskíði*).

langlokum *f. pl.* (dat.) 'with late conclusions' or 'with long enclosings'? 14/1 textual n. (cf. *FoGT* 136; see 14/1 n.).

langr *a.* long 0/29, 23/4 (of ships), 36/2, 71/8 (with *hesta*, i.e. ships); comp. 2/13; n. as adv. further 8/24, 30.

láta (lét) *sv.* with acc. and inf., cause s-thing to do s-thing, let (make) s-thing (s-one) do s-thing 6/1, 7/5, 11/3, 21/7, 42/5, 75/1, 77/1; with inf. make [people] do s-thing 59/3 ('lets shields be broken'); with acc. and pp. cause s-thing to be s-thing: *lætr ofsa bannat* 1/1, *lætr fal framðan* 18/5, 18, *létu mik virðan* 67/6 (the subject is *jǫfrar*), *lætr hǫpp hrest* 93/1, *lét ek jǫfur sótt* 93/6; with pp. of impers. vb. *lætr ósvipt húnskript* 78/7; with acc. and a. 66/5; declare, consider, judge s-thing [to be] s-thing 8/31.

laukr *m.* leek; in kenning for sword (part of kenning for shield): *Mistar l.* 85/2; = mast 77/6 (instr.).

laun (1) *f.* concealment 18/10, 15.

laun (2) *n. pl.* reward, repayment 26/6.

lauss *a.* free, unattached 10/4; *laus vísa* separate verse not part of a sequence or long poem 53/13.

léa, ljá (ð) *wv.* grant (*e-m e-s*); *l. tiggja jarðar* 12/3.

legi *see* **lǫgr**.

leggja (lagða) *wv.* lay; *l. við* add to, increase 27/15.

leggr *m.* leg, arm, limb; acc. pl. (obj. of *brýtr*) 61/6.

leið *f.* way; sea 34/3; in kenning for arms, *vala l.* 48/8 (obj. of *venr*).

leiða (dd) *wv.* lead; *vera leiddr af* be dependent on 15/9; impers. (?) *leiðir e-t af* s-thing belongs in sense with 27/10, 39/9. (The word is used to mean 'pronounce' in *SGT* 66.)

leika (lék) *sv.* play; with acc., play with or in (move to and fro in) 36/5; *l. við* move to and fro against 87/5.

leita (að) *wv.* with gen. seek 6/6, 16/16 (impers. pass., sc. *orðanna*); *l. e-s at þar er* seek something (there) where 6/12.

lemja (lamða) *wv.* strike, beat; impers. *lemr e-t (á e-u)* 38/2 ('waves are struck').

lengð *f.* length 7/12, 33/10 (cf. **orðalengð**).

lengja (ngð) *wv.* lengthen 8/13.

lengr *adv. comp.* for longer, i.e. for the (foreseeable) future 26/8.

lengri *a. comp.* see **langr**.

lest *f.* cargo; *lesta hestr* = ship 44/17.

lesta (st) *wv.* damage, destroy; *l. bil*, i.e. be decisive, act quickly 37/4.

léttr *a.* light; n. as adv., easily 23/6.

leyfa (fð) *wv.* praise; pp. honoured 71/7, 79/1.

leyfi *n.* permission, licence 0/7, 1/24, 6/22, 8/15, 17, 18, 24, 25 (cf. *TGT* 91, 107; *leyfiligr*, *FoGT* 130).

lið (1) *n.* following, troops 1/2; with gen., troop, company 23/5, 47/2; men 43/6.

lið (2) *n.* vessel 44/18.

líða (leið) *sv.* move 22/6, 33/4; pass (of time) 23/2; pp. having slid, having been placed (*at e-u* on to s-thing) 45/7 (with *glóðir* = rings), past 94/6.

liðhending *f.* 'helping rhyme' (rhyme and alliteration combined in the same syllable) 8/26, 58/12.

liðhendr *a.* 'help-rhymed', when rhyme and alliteration fall on the same syllable 41/10.

liðhendur *f. pl.* 'helping rhymes' 40/10, 41/9, 52/10, 53/9; dat. pl. 32/1 textual n. (error for *riðhendum*). This name is given to the verse-forms exemplified in both stt. 41 and 53.

liðr *m.* limb; *at lið* on to limb (collective) 45/7; in kenning for gold, *liðar sker* 46/2.

liðsmaðr *m.* member of a troop 4/15.

líf *n.* life 4/14, 33/1 (gen. with *rán*).

lifa (ð) *wv.* live 4/4, 96/5.

liggja (lá) *sv.* lie 10/8, 35/11; cf. **lá**.

limgarmr *m.* branch-hound, i.e. enemy of branches or trees, kenning for wind or storm 78/4 (subject of *brýtr*).

limr *m.* limb 10/7 (collective); *l. axla* = arm 2/2 (obj. of *hylr*).

lind *f.* (lime-wood) shield or spear 9/2 (dat.), 10/3 (acc.); in kenning for battle 32/2.

linnr *m.* snake; in kenning for sword, *l. senna sverða* 'snake of quarrel of swords' 6/5.

linr *a.* soft; i.e. weak, unaccented? 0/29 n. In *TGT* 39, 54 the word refers to lack of aspiration. Cf. *GT Prologue* 154.

lita (að) *wv*. colour, redden (with blood, by fighting battles) 4/5, 8/4, 96/4; md. for passive 9/8.

líta (leit) *sv*. see, look at 8/6; with acc. and inf. 75/5.

lítill *a*. small; n. as subst. s-thing small 27/8 (litotes: 'for nothing' or 'for small reward').

lítt *adv*. little, not much, i.e. not at all 24/2, 64/1, 92/4.

ljóðaháttr *m*. 'song-form' 99/9. (See Wessén 1915, 129–34.)

ljóma (að) *wv*. shine 22/5.

ljóss *a*. bright 69/7 (with *elds*), 97/1 (with *lofi*); clear, obvious ('of obvious meaning'?) 17/16; n. as adv. 17/14, 24. Cf. **ofljóss**.

ljósta (laust) *sv*. strike; impers. with dat. (*Mistar regni*) there strikes down 62/7 n.

ljótr *a*. ugly 22/4 (with *lægir*), 93/5 (with *haf*).

loðbrók *f*. 'shaggy-breeches', nickname of the viking Ragnarr (see *Ragnars saga loðbrókar*, ch. 3; 1906–8, 118) 53/14.

lof *n*. praise, eulogy 31/4, 68/2, 97/2, 102/8 (parallel to *fold*: 'before the ruler's praise (cease, be forgotten)'); poem of praise 80/1, 96/7 (subj. of *mun lifa*, = *þat* 96/5, and of *ort er* 96/1?). See Fidjestøl (1982), 253–4.

lofa (að) *wv*. praise 93/7.

lofðungr *m*. king 26/7 (cf. *Skáldskaparmál*, ch. 64).

lofkvæði *n*. poem of praise, encomium 67/15.

lofun *f*. praise, eulogy, poem of praise 69/8 (subj. of *skal vaða*).

logi *m*. flame 17/17; in kenning for gold, *l. ǫldu* 17/4, 17, 29.

lok *n*. end 102/4.

lúka (lauk) *sv*. with dat. end, complete 13/11 (cf. **fylla**); impers. with dat. 11/9, 14/9, 70/13 (inf. with *skal*); md. (= *endask* 71/11?) 23/10 n., 36/10, 78/10; *lúkask á, í* be followed by? contain? 32/10, 73/10 (in these two phrases the reference seems to be to the endings (or second syllables) of words containing the same sounds; see 16/12–13 n.). Cf. *FoGT* 137–8.

lund *f*. manner; *á þá l. at* in such a way that 53/9.

lundr *m*. grove; tree; in kenning for warrior: *hjarar l.* 60/8 (gen. pl. with *sund* or *unda* or *fólk*). Cf. **steykkvilundr**.

lung *n*. ship ('longship') 21/1, 34/8 (dat. of respect).

lúta (laut) *sv*. with dat. bow to 13/3.

lýðr *m*. people, troop 21/1, 23/5 (gen. pl. 'of men'), 26/7 (with *ráð*, 'over men'), 77/5.

lypta (pt) *wv*. with dat. lift, raise 97/1; md. (for passive) 52/2 (subject is *skǫpt*), 77/5.

lýsa (t) *wv*. illumine; make famous, shed glory on 20/8 (obj. is *vísa*; 'show what sort of person he is'?—or the verb may be intrans., 'shine', and *vísa* gen. with *iðjur*).

lýsheimr *m*. pollack-home, i.e. sea, in kenning for gold 22/5.

lýslóð *f*. pollack-track, i.e. sea, in kenning for gold (rings) 45/8.

lýti *n.* blemish, deformity 10/7.
læ *n.* destruction; fraud 17/20.
læsa (st) *wv.* lock, secure (*e-t e-u* s-thing with s-thing) 79/1.
lœgir *m.* ocean ('lier, calm one') 22/3.
lǫgr *m.* sea 16/6 (the phrase belongs with *þryngr*), 20/4, 23/3, 44/17, 78/3 (dat., obj. of *skýtr*); in kennings for gold, *hagbál lagar* 44/6, *eldr lagar* 69/7; for blood, *benja legi* (dat. sg. with *venja*) 60/6.
lǫgstígr *m.* sea-path (in kenning for ship) 22/3.
lǫngum *adv.* (dat. pl. of **langr**) for a long time 21/6.

maðr *m.* man 91/7 (obj. of *hittir*), 92/7 (with *mildara*, after *né*); indefinite subject of *viti* 95/8, of *skala* 100/5.
máhlíðingr *a.* of Mávahlíð on Snæfellsnes in western Iceland 8/21.
mál (1) *n.* (suitable) time, opportunity 75/4 (subject of *falla*).
mál (2) *n.* speech; in kenning for mead, *máls heilsa* 25/5; discourse, language 4/15; *rétt at máli* correct in usage 15/12; meaning, content 0/28 (as opposed to sound; see Konráð Gíslason 1875, 98), 1/47, 48, 6/20, 8/45, 9/11, 17/26, 18/14 (cf. *Skáldskaparmál*, ch. 1, *SGT* 52); subject, way of talking, reference 6/14; statement, utterance, sentence 9/9, 10/9, 11/9, 13/10, 12, 14/9, 70/13 (cf. *TGT* 99, *FoGT* 136–7); *vera sér um mál* comprise, make up a separate statement 12/12, 13/10, 14/10, 27/11 (cf. *FoGT* 137); affairs, lawsuits 84/5 (if *hinn er mál metr* is a description of *milding*; if it is the subject of *getr* it means 'one who is a judge of matters, i.e. a discerning man').
mála *f.* woman-friend; *Heðins m.* = Hildr as personification of battle 49/5; in kenning for land (of Norway): *mála úlfs bága*, i.e. Óðinn's wife Jǫrð 3/2 (obj. of *verr*).
málaháttr *m.* 'speeches-form' 94/9. (See Wessén 1915, 129–34.)
málfylling *f.* particle (unstressed grammatical word) 1/23.
málmr *m.* metal; in kenning for battle 52/8.
málmskúr *f.* metal-shower; *málmskúrar dynr* = battle 39/4.
málrúnar *f. pl.* speech-runes (runes as alphabet, phonetic runes) 1/43. Cf. *málstafr* in *FGT* 222, when it means (runic) letter; in *SGT* it means consonant.
málsgrein *f.* distinction of meaning (content; as opposed to sound) 0/15 textual n., 0/26, 27; distinction of language 67/12. Cf. *FGT* 224–6, *Skáldskaparmál*, ch. 1; in *TGT* 37, 56–8, 60, 94, 98, 111, 112, *FoGT* 144 the word appears to mean sentence, in *TGT* 72, 79 = syntax, in *TGT* 61–2 = (use of) language.
málsorð *n.* word 8/27, 15/10, 16/9, 24/9 (here the reference is to the root syllables of words only), 58/9, 62/11; *fyrir m.* at the beginning of a word 82/11 n. (cf. *fyrir samstǫfun* 1/14).
máltak *n.* (selection of) meaning, significance 39/9 (= *orðtak*?); *ǫnnur*

máltǫk different ordering of phrase, expression or signification 18/14. Cf. *Skáldskaparmál*, ch. 1: 'choice of language'.

mannbaldr *m.* outstanding person 36/6 (complement of *reyndisk*). Cf. Baldr in Index.

manndýrðir *f. pl.* virtues, glorious achievements 13/1, 44/4.

marblakkr *m.* sea-steed, i.e. ship 46/8 (*see* **blakkr**).

margdýrr *a.* most valuable 29/6.

margr *a.* many 30/6, 70/14, 90/1; i.e. frequent 67/15; many a 56/4, 66/5; n. *mart* 70/2 (with *lag*).

marr *m.* sea 21/8; *verpa á aldinn mar*, i.e. waste, throw away 67/7.

máttr *m.* might; *of mátt* with might, of mighty proportions 89/6.

meginbára *f.* mighty wave; in kenning for blood, *m. sára* 60/2 (subject of *skylr*).

meginskíð *n.* mighty rod or plank, in kenning for spear or sword: *m. fleina lands* 65/6 (dat. with *rendi*).

meginund *f.* mighty wound 63/8 (obj. of *stærði*).

meiðr *m.* tree, in kenning for men or warriors: *meiðar sævar rǫðuls* trees of the sun of the sea, i.e. of gold 17/8 (dat. of respect), 24 (also 1st pers. pl. of *meiða* wv. 'injure'), 32.

meir *adv.* further 68/5 ('I shall continue to . . .').

meizla *f.* (= *meiðsla*) injury, mutilation 17/25.

meldr *m.* meal, the produce of a mill; in kenning for gold, *Fenju m.* 43/8 (cf. *Skáldskaparmál*, ch. 43).

mella *f.* trollwife; in kenning for Þórr, *mellu dólgr* 3/8.

men *n.* neck-ring, gold collar 45/3 (acc. with *fyrir*); in kennings for ruler, *menja fættir* 45/3, *mens stiklir* 60/6.

menglǫtuðr *m.* neck-ring destroyer, generous prince (indefinite) 95/7.

menstiklir *m.* neck-ring thrower, generous prince, = Skúli 44/3 (subject of *vann*; cf. **stiklir**).

menstríðir *m.* neck-ring afflicter, enemy, i.e. generous man 3/7 (vocative; = Hákon).

merki *n.* banner, standard 33/4, 52/7, 64/4.

merkir *m.* marker; in kenning for warrior: *m. blóðsvara* he who marks birds of prey (with blood, by providing dead bodies for them in battle) 92/8 (generic sg., 'no warrior will find').

meta (mat) *sv.* measure, judge, adjudge 84/5 (see **mál (2)**).

miða (að) *wv.* impers. there is movement (i.e. the water moves) 38/8.

miðr *a.* middle of 1/41, 97/11.

mikill *a.* great 4/14, 14/4, 39/2 (with *hjálmar*); dat. sg. n. *myklu* as adv. with comp. much 8/24, with sup. by far 94/8; comp. *meiri* greater 49/10, 59/11; sup. *mestr* (the) greatest, very great, maximum 17/33, 36, 19/9, 12, 37/4 (with *gjald*; or with *verð*, see under **gjald** and **verðr**), 50/9, 60/9, 88/8, 93/2.

mikla *adv.* greatly, very 44/3 (with *vensk*; or a., acc. pl. with *frama*? or emend to *miklar*, a., f. pl. with *manndýrðir*).

mildingr *m.* kind, generous ruler; = Hákon 25/5, = Skúli 84/6 (obj. of *getr*), 95/1 (dat. of respect?—see **muna**).

mildr *a.* generous, kind (always of rulers) 10/6, 68/3; *m. e-s* generous with, unsparing of s-thing, in description of ruler, *skjaldbraks m.* 28/3; comp. with *mann* 92/7 (obj. of *finnrat*).

milli *prep.* with gen., between; *m. Gandvíkr ok Elfar* 1/6.

minn *poss. a.* my; n. *mitt* 70/1 (with *lag*; 'many a verse-form of mine').

minnask (t) *wv. md.* remember; *m. á e-t* call s-thing to mind, recall, mention s-thing 31/2.

minni (1) *n. pl.* memorial, what is remembered or serves to remind; *at minnum e-m* as a memorial for s-one('s benefit) 67/1; tradition, inherited statement, proverb 13/12.

minni (2) *a. comp.* lesser 19/14, 20/10, 42/11, 43/9; sup. *minztr* least 19/12, 22/9, 81/10, 87/9.

missa (t) *wv.* impers. with gen., there is lack of, s-thing is lacking 58/10.

mjúkr *a.* smooth 65/14.

mjǫðr *m.* mead 24/3, 25/6, 91/2 (obj. of *þiggja*).

mjǫk *adv.* very 36/2, 84/3 (with *rausnsamr*); to a great extent 39/10; greatly, very much 26/5, 33/2, 87/3 (with *þekkr*); often, a great deal 6/11.

móðir *f.* mother; in kenning for land (Norway): *mellu dólgs m.* i.e. Þórr's mother Jǫrð 3/7.

móðsefi *m.* mood-thought; in kenning for breast 50/2.

morðaukinn *a.* (*pp.*) made great, famous by killing 49/7 (with *mæki*).

morðfár *n.* 'killing-danger', i.e. sword; in kenning for warrior (Þórarinn máhlíðingr): *morðfárs myrðir* sword-destroyer 8/23.

morðflýtir *m.* one who urges, hastens or promotes killing, war-leader (Skúli) 39/3.

mót (1) *n.* manner, kind 53/9 ('in twofold fashion'), 79/12 ('all have the same characteristic').

mót (2) *n.* meeting; pl. in kenning for battles, *m. málms* 52/7.

mót (3) *prep.* with dat., towards, against 20/4 (and by word-play adv., 'back (towards)'); *í móti* contrary to, the opposite of 17/22.

muna (man, munða) *pret.-pres. vb.* remember, keep in mind 15/3; *m. e-m* (dat. of respect?) *e-t* remember s-one for s-thing, remember s-thing as due to s-one 29/1 (here the dat. may go with *heimsvistir*), 95/1.

mundr *m.* bridal, wedding-gift, gift from husband at betrothal 49/8 (in apposition to *mæki*, 'as a wedding-gift', i.e. Skúli wages war—being betrothed to Hildr, i.e. battle).

munnroði *m.* mouth-reddening; *auka munnroða* add redness to the (raven's) mouth (with blood, by offering carrion as a result of fighting battles) 5/6.

munnvǫrp *n. pl.* 'mouth-throwings', improvisation 65/15.
myklu *see* **mikill**.
myrðir *m.* destroyer, enemy; in kenning for warrior (Þórarinn máhlíðingr): *morðfárs m.* 8/22.
mækir *m.* sword 49/7.
mæla (t) *wv.* say, express 4/10, 17/13, 24.
mælingr *m.* niggardly person 40/2 (dat. with *ferri*).
mær *f.* (gen. **meyjar**) daughter; *Hǫgna m.* = Hildr, personification of battle (in kenning for shield) 49/3.
mærð *f.* glory 13/1 (gen. obj. of *fá*); poem of praise, encomium, eulogy 68/4, 85/1, 92/2, 95/7 (obj. of *viti*), 97/4.
mæti *n. pl.* objects of value 46/7 (obj. of *þakka*, 'for objects . . .').
mætr *a.* noble 13/2, 46/7; splendid 28/8 (with *stórlæti*); fitting, worthy, honourable 45/4.
mœða (dd) *wv.* make tired 23/5.
mœta (tt) *wv.* with dat. meet, face 39/3, 64/1, 65/3, 77/8, 91/5.
mœtir *m.* meeter, one who faces; in kenning for warrior, *oddbraks m.* 70/1 (indefinite).

ná (ð) *wv. aux.* with inf., get to, manage, be able to do s-thing 8/6, 22/5 etc.; with *at* and inf. 14/5, 24/2.
naðr *m.* snake; in kenning for sword, *n. sóknar* 6/1; for winter 83/1.
nafn *n.* name 8/39, 46.
náhendr *a.* 'close-rhymed', with rhyme in adjacent syllables (cf. **riðhendr**) 74/11, 92/10; n. as subst. 74/11.
náttúra *f.* nature 6/11.
nauðr *f.* necessity; in kenning for arm, compeller, forcer: *boga n.* 48/2.
ne *neg. adv.* not 19/5, 31/6.
né *conj.* nor 67/5, 92/7; and not 80/3.
nefna (fnd) *wv.* name 4/16, 8/38.
nema (1) *conj.* unless, except; *n. svá at* if it were not that 56/3; i.e. until 96/6.
nema (2) (nam) *sv.* touch, strike 61/3; with inf. begin to (as verse-filler: 'did') 92/4.
niðr *m.* descendant; relative; *skjǫldungs n.* = Skúli 52/1 (half-brother of King Ingi, see 33/5 n.). Skúli was not of royal blood (unless very distantly on the female side, cf. *Bǫglunga sǫgur* (1988), ii. 26–7; *Hákonar saga*, ch. 199, and ch. 242, 'þeirra sem eigi váru af sjálfri langfeðgatǫlu konunganna'), so *niðr* cannot mean 'descendant' here, unless Snorri is grossly flattering him; but it does occasionally mean 'brother' (see *LP*, *niðr* 1). The alternative is to take *niðr* as adv. 'down', *skjǫldungs* with *skúrum* and *sær* as impers. for pass.
niðri *adv.* below, underneath 38/8.
niðrlag *n.* end, conclusion 70/16.

njóta (naut) *sv.* with gen., enjoy, be possessor of, derive benefit from 72/7; imp. (optative) 3/7, 30/5; subjunc. ('let them enjoy') 102/1.

nýbitinn *a.* (*pp.*) newly bitten, just pierced 56/6.

nýgjǫrving *f.* extension of meaning, i.e. metaphor; in pl. extended metaphor, allegory 1/54, 5/12, 6/9, 12, 13, 20. The word refers to giving new meanings to words, not to original phrasing or neologisms. Cf. *Skáldskaparmál*, chs. 33, 50, 69; *TGT* 80.

nykrat *a.* (*pp.*) n. as subst., (made) monstrous, monstrosity 6/16. Cf. *TGT* 80, *FoGT* 131.

nýr *a.* new 39/11, 72/10.

nýta (tt) *wv.* make use of, get benefit from 44/18 (i.e. 'we travel fast', cf. **njóta**); gain an effect, use effectively 8/28 (or 'be acceptable, allow [it]'?). Cf. **ónýtr** and *FoGT* 147.

nýtr *a.* beneficial (*e-m* to s-one) 30/7.

næst *adv.* next 28/5; with dat., next after 65/11; *þar næst* and then 4/14.

næstr *a. sup.* closest (*e-u* to s-thing) 99/8 (i.e. experiencing the greatest *happ*); next, adjacent to 28/13.

óbreyttr *a.* (*pp.*) unchanged; ordinary, without variation 64/9.

óbrugðinn *a.* (*pp.*) unchanged, without deviation, without being departed from 6/18.

óð *see* **vaða**.

oddbrak *n.* point-clash, -crack, i.e. battle; *oddbraks mœtir* = warrior 70/4.

oddhending *f.* front-rhyme, rhyme which comes at the beginning of a line 1/40, 7/9, 8/1 textual n., 41/10–11.

oddhendr *a.* having a rhyme-syllable at the beginning of the line 76/11; n. as subst. 7/1 textual n.

oddr *m.* point (of weapon) 79/6 (instr.); pl. by synecdoche = weapons 54/7 (obj. of *rýðr*), 63/5 (dat. obj. of *rendi*); = arrow in kenning for shield 31/5.

oddviti *m.* leader (= Skúli) 56/5, 59/8 (with gen. pl.), 66/8, 88/2 (with gen. sg.).

óðharðr *a.* mighty hard 5/1, 10 (complement).

óðr *m.* poetry 31/8.

of (1) *prep.* with acc. and dat., over, around; with acc. concerning 45/4, 67/3.

of (2) pleonastic particle with verbs 56/3.

ofljóss *a.* containing *ofljóst* (lit. 'excessively clear'), punning, using word-play 17/26, 20/9; n. as adv. 18/13. Cf. *Skáldskaparmál*, ch. 74, *TGT* 66, 89, 172.

ofrhugaðr *a.* most bold 5/4.

ófriðr *m.* lack of peace, hostility (-ies), warfare 17/31.

ofsi *m.* arrogance, violence 1/4 (obj. of *bannat*).

ógn *f.* threat, attack, battle 58/3; in kenning for warriors 62/1; in kenning for winter, *alla naðrs ógn* throughout the adder's terror (acc. of time) 83/1 (cf. *Skáldskaparmál*, ch. 29).
ógnflýtir *m.* attack-, battle-hastener or -promoter (i.e. Skúli) 42/7.
ógnrakkr *a.* attack-, battle-bold 42/2 (with *snerpir*).
ógnsvellir *m.* battle-increaser, 'war-sweller' (i.e. Skúli) 39/7.
ógnþorinn *a.* (*pp.*) battle-daring 5/8.
óhneppr *a.* not obscure, not shabby, not unremarkable (*hneppr*: barely sufficient), i.e. successful 70/6 (with *skrautfara*).
ójafn *a.* not equal, not equivalent, not the same thing 17/25.
ok *conj.* and; links *randgarði* and *bláskíðum* 79/3, links *átti ráða* and *var faldinn* 15/2; as well as 12/1, 10.
ókveðinn *a.* (*pp.*) unrecited, not composed (in), not used in poetry 70/3 ('many of my verse-forms have never been used before').
ólestr *a.* (*pp.*) unblemished, complete, pure 93/4.
ólíkastr *a. sup.* most dissimilar, most contrary 16/12.
ólíkr *a.* dissimilar, different, opposite 17/11.
ólítill *a.* no small (litotes) 55/6.
ónýtr *a.* unacceptable or not allowed? ineffective? worthless? 8/31 (cf. **nýta**; *FoGT* 147).
orð *n.* (1) word (= *málsorð*) 1/32, 36, 4/9, 16/13; *með fullu orði* comprising a complete word 13/10, with a complete word 27/10; in kenning for teeth 87/8; pl. statement or phrase? 17/16. (2) line (= *vísuorð*) 1/40, 41, 4/13, 7/12, 8/13; = word or line 80/10; *með orðum* = either 'in words' (i.e. it is not a metrical variant) or 'in lines' (i.e. it is similar in structure, varied in *mál*) 40/9 n. (3) reputation, renown, fame 45/4, 84/1.
orðalengð *f.* length of line 2/9, 9/9, 23/12, 15, 27/15.
orðfimi *f.* verbal skill or dexterity, agility 6/21.
orðfjǫlði *m.* store of words, vocabulary 6/19.
orðrómr *m.* renown 14/8, 82/4 (obj. of *á*).
orðskviðuháttr *m.* proverb-form 25/9, 26/1 textual n. (contains proverbs or gnomic statements in syllables 2–6 of the b-lines).
orðtak *n.* turn of phrase, arrangement of words (cf. **máltak**) 8/50, 17/26; expression 13/12, 34/10; language, choice of words 16/12, word (chosen) 72/10. Cf. *TGT* 45, 101, 109, 115. In *Gylfaginning* the word means 'saying'.
ormr *m.* snake 6/9, 10, 15, 94/5 (= Fáfnir, dat. with *veitti*); in kenning for Þórr 3/5 (= *miðgarðsormr*, cf. *Gylfaginning*, ch. 51); in kenning for sword, *o. vals* 6/7, for winter, *orms galli* 83/5.
orpit *see* **verpa**.
ort- *see* **yrkja**.
orrosta *f.* battle 2/12, 18/16.

ósiðr *m.* bad custom, immoral practice 17/22 (contrasted with *siðr* 'custom'), 30.

ósléttr *a.* uneven, not level 17/12.

ósviptr *a.* (*pp.*) unreefed (*svipta* wv. with dat., to reef); *lætr e-u ósvipt* cause s-thing to be unreefed, i.e. he does not have (the sail) reefed, sails without fear of the high wind (cf. *Hálfs saga* 1981, 178/35 [read *hálsa*], *ÓH* 666) 78/8.

ótal *n.* a countless number (of) 88/4.

otrgjǫld *n. pl.* otter-payment, compensation for death of otter, i.e. gold (cf. *Skáldskaparmál*, ch. 39) 41/2.

ótt *adv.* (n. of *óðr* a.) fast 6/3.

ótvistr *a.* not unhappy, (very) enjoyable 29/2.

óvarr *a.* unwary, unhesitant, brave 80/7 (with *bǫrum*).

óx *see* **vaxa**.

prýðir *m.* adorner, one who dispenses honour to (with gen.); *jarla p.* = King Hákon 27/7 n.

ráð *n.* (power to) rule (with gen., over) 26/7.

ráða (réð) *sv.* with dat. rule 1/5, 14/1 (the obj. is *konungdómi* 14/8), 15/4 (with *þeim er*); abs. 15/7 (or 'undertake, perform', with *stórt* as obj.); determine, govern 1/9, 13, 29/9, 30/9 (abs., 'be the determining factor'), 54/10, 82/12; plan, decide 17/31; abs. be the cause 33/2.

ramr *a.* strong, powerful 38/6.

ramsnákr *m.* strong snake; in kenning for sword, *rógs r.* 6/4.

rán *f.* plunder, robbing 33/1; robbery 17/22, 66/7.

randgarðr *m.* shield-enclosure, shield-wall 79/4 (instr.).

rangbarmr *m.* curved side or rib-edge (i.e. side of ship; *rǫng* f. 'rib') 78/4 (subject of *skýtr*).

rangr *a.* wrong, incorrect, contrary to the rule 1/19, 44/13, 70/11, 75/9, 77/9, 82/11.

ránhegnir *m.* punisher (or preventer, checker) of plundering 26/3.

ránsiðr *m.* practice of plundering, plundering behaviour 17/7, 22, 32.

raska (að) *wv.* disturb; impers. with dat. (= pass.) 9/6.

ráskegg *n.* yard-beard, beard of ship's yard-arms, i.e. sail 78/6 (obj. of *rekr*).

rauðr *a.* red; of gold 37/6 (with *auð*), 46/6 (with *sæfuna*), 48/4; n. as subst. *falda rauðu* put on red (caps) 63/2 (see **falda**).

rauðsylgr *m.* red drink; in kenning for blood, *bens r.* 56/2 (dat. [instr.] with *ali*).

raukn *n.* draught animal (ox or horse); in kenning for ship, *kjalar r.* 77/6.

raun *f.* tried qualities, experience 26/6 (dat.); pl., reality 33/1 (*skapat at* *raunum* 'made into a reality'; or *at raunum* 'in truth, in accordance with experience, as they truly discovered'?).

raungóðr *a.* proved good, of tried goodness 75/7.

rausn *f.* magnificence (of behaviour, i.e. hospitality) 89/6.

rausnsamr *a.* splendid in one's way of life, living in great style (referring to hospitality) 84/4 (complement).

refhvarf *n.* 'fox-turn', 'fox-trick' (or 'fox-lair'?), antithesis (the name presumably relates to the word-play, and refers to the cunning or deceptive meanings or to the sudden shifts of sense in the antitheses) 20/9; usually pl. 17/35, 18/9, 19/10, 14 (*ein*: one pair of), 21/9, 22/9, 23/9; as the name of a verse-form 16/11, 17/1 textual n., 17/34 (sg. 17/34 textual n.), 18/1 textual n., 19/1 textual n.; *in mestu refhvǫrf* 17/33, 36, 19/9, 12; *in minni refhvǫrf* 19/14, (20/10); *in minztu refhvǫrf* 22/9; *refhvarfa bróðir* 22/10.

refhvarfaháttr *m.* *refhvǫrf*-form 18/21.

refhvǫrfmæltr *a.* (*pp.*) expressed in *refhvǫrf* 17/15.

refsa (að) *wv.* punish (*e-m e-t* s-one for s-thing) 66/7.

regg *n.* a kind of ship 34/4 (instr.).

regn *n.* rain; in kenning for battle, *Mistar r.* 62/8 (obj. of *lýstr*).

reið *f.* chariot, carriage, where s-thing is carried, resting place; in kenning for arms: *heiðis reiðir* 48/6 (obj. of *býr*).

reiða (dd) *wv.* brandish 2/8.

reiðmálmr *m.* 'riding-metal', metal carried on horseback; in kenning for gold, *r. Gnitaheiðar* 41/6 (cf. *Skáldskaparmál*, ch. 40).

reiðr *a.* angry 17/8 (with *ræsir* or *fylkir*), 17/24, 66/8.

rein *f.* ridge or strip of land; *Røkkva reinar* = sea in kenning for ships 73/4.

reip *n.* rope 34/6 (instr. with *harðsveipaðastan*).

reisa (t) *wv.* raise; md. for pass. 36/1.

reitr *m.* (acc. pl. *reitu*) a strip of land; in kenning for arm (where the hawk sits): *vals r.* 42/3 (*firum* is dat. of advantage).

reka (rak) *sv.* drive; of a horse or ship 20/6 (also by word-play 3rd pers. sg. of *rekja* wv., 'spread out, unfold'), 34/5, 78/5 (of the sail); drive away (or fulfil? avenge?—but then the obj. should be gen.) 17/3, 15, 30; drive back 18/12 (pp.); of metal-working, hammer: pp., inlaid (with *fal*) 18/6, 18; extend (a metaphor), use extended kennings (i.e. with more than two determinants) 1/53, 8/29; pp. 2/11, 13, 8/30. Cf. *Skáldskaparmál*, ch. 50. In *Háttalykill*, *rekit* is the name of the verse-form corresponding to *Háttatal*, st. 90 (8-syllable *runhenda*).

rekkja (kkt) *wv.* embolden, encourage, strengthen, urge on 1/2; try the strength of? enliven, make dance? (cf. *NN* 2176) 22/7.

rekkr *m.* man, warrior 26/5, 38/3, 63/3.

remma (ð) *wv.* cause to be strong, increase (s-thing) in strength or intensity 57/7. Cf. **gilda**, **stœra**.

remmi-Týr *m.* god who encourages, increases the intensity of s-thing; in kenning for ruler (Hákon): *rógleiks r.* encourager of war 14/6.

renna (1) (rann) *sv.* run, flee 33/8; pp. flooded, soaked (*e-u*) 32/4.

renna (2) (nd) *wv.* cause (*e-u* s-thing) to run over or through (*e-t* s-thing) 63/5, 65/5, 101/5. Cf. **hleypa**.

rétthendr *a.* consistently rhymed; n. as subst. as name of a verse-form 41/13.

réttr *a.* straight 6/3; direct 20/5; correct, normal 0/11, 12, 1/24, 4/15, 6/18, 19, 23, 34/11, 43/11, 68/11, 70/13, 76/10, 77/12; *r. at máli* correctly constructed 15/11; proper 17/26, i.e. good 23/6 (of *ǫl*; with the reading *in* it would refer to *hǫll*); of *runhenda*, having the same rhyme throughout the stanza (= *full runhending* 83/9) 79/13, 86/9; n. as adv. straight 63/3, correctly 41/10, 43/9, i.e. literally 4/10, appropriately 6/9, with the same rhyme thoughout the stanza 89/9; *rétt at stǫfum* with normal alliteration 76/10.

reyna (d) *wv.* try, put to the test 27/3; md., turn out to be, prove (*e-m* to s-one) 36/5.

ríða (reið) *sv.* ride; of gold, fly, be scattered, be distributed 41/5.

riðhendr *a.* 'rocking-rhymed', with rhyme-syllables close together at the end of the line; n. as subst. 55/10, 56/10. Cf. *TGT* 98.

riðhendur *f. pl.* 'rocking rhymes', rhymes close together at the ends of lines 31/12, 45/12.

ríki *n.* kingdom, realm 14/5 (obj. of *stýra*), 18/17.

ríkr *a.* powerful 27/4 (with *ræsis*).

rimma *f.* tumult, battle; in kenning for swords 57/7.

rísa (reis) *sv.* rise 21/2.

rísta (reist) *sv.* cut, incise, carve 35/1, 101/4 (obj. is *straum*).

rít *f.* (engraved) shield 8/6 (pl., obj. of *líta*), 54/4, 73/3 (instr.).

rita (að) *wv.* write 2/10, 8/48; write in, exemplify 11/9, 23/11, 27/12, 13, 51/9, 13, 53/11, 54/13, 61/10.

rjóða (rauð) *sv.* redden (with blood, in battle) 7/7, *r. í e-u* 54/7; pp. *roðinn* 4/15, 57/2, 63/4 (painted red?), 64/4 (perhaps here = golden? cf. **rauðr**); of reddening of the lips of wolves by the dead bodies provided for them in battle 11/8, 96/2.

rjóðr *m.* reddener (with blood); in kenning for warrior (= Skúli): *hjǫrs r.* 41/5.

rjóðvendill *m.* red(dening) rod, in kenning for sword: *r. randa* 13/5.

rjúfa (rauf) *sv.* rip, tear 10/2, 50/1.

ró = *eru* 99/1.

róa (røra) *sv.* row 72/2.

roðinn *a.* (*pp.*) see **rjóða**.

roðna (að) *wv.* become red (with blood) 9/1, 56/8 (i.e. they cause men to fall in battle and become prey to wolves).

rofna (að) *wv.* be broken 18/15.
róg *n.* slander, strife, battle, in kenning for sword 6/3.
rógálfr *m.* 'hostility-elf', i.e. war-leader, ruler (Skúli) 75/8 (subj. of *lítr*).
rógleikr *m.* strife-game, i.e. war, in kenning for ruler 14/5.
rúm *n.* place; position, bench in a ship 21/6 (also by word-play n. pl. of *rúmr* a., 'spacious').
rúna *f.* confidante, wife; in kenning for land (Norway): *Míms vinar r.* = Óðinn's wife Jǫrð 3/4.
rúnar *f. pl.* runes, runic alphabet 1/43.
runhenda *f.* end-rhyme (to *runi* m. 'running, flowing' or *runa* f. 'string, row, list') 79/11, 13 (*rétt r.* a stanza with the same rhyme in each line, = *full r.*), 88/9, 90/9, 91/9, 92/9 (*full r.*); *minni r.* (with four lines rhyming together) 80/12, 86/11, 89/10, 92/9; *minzta r.* (with lines rhyming in pairs) 81/11, (87/9).
runhending *f.* end-rhyme (end-rhyming form) 80/10, 83/9 (*full r.*), 84/9, 86/9 (*rétt r.* a rhyming form with the same rhyme throughout the stanza), 88/11.
runhendr *a.* having end-rhyme 79/12, 82/13 (m. pl. with *hættir* understood), 85/11, 89/9 (*rétt r.* 'having the same rhyme throughout the stanza'), 91/10; n. as subst. as name of a form 80/11.
ryðja (rudda) *wv.* clear; md. for pass., be cleansed (of rebels) 64/5.
ræsa (t) *wv.* (cause to) move (quickly) 17/23; make flow (blood) 7/4.
ræsir *m.* mover, leader (of battle or of men in battle), ruler; = Hákon 17/7 (cf. **ræsa**), 32, 26/5, 27/3; = Skúli 73/1, 91/2, 96/1; *r. Þrænda*, i.e. Skúli 64/5.
rœði *n.* oar 75/7 (pl., subject of *skjálfa*); in kenning for tongue, *tǫlu r.* 81/4.
rœki-Njǫrðr *m.* Njǫrðr (a god) who uses or cultivates the use of; in kenning for warrior (i.e. Hákon): *r. rjóðvendils randa* sword-user 13/6.
rǫðull *m.* sun; in kenning for gold (part of kenning for warriors): *rǫðuls sævar meiðar* 'trees of sun of sea' 17/6 (the emendation from *rǫðul* seems established by 17/32, though it would be possible to take *rǫðul-sævar meiðar* = *rǫðul-meiðar sævar* with the same meaning, cf. **vandbaugr** and **vandbaugskaði** and note to 15/1–2), 17/20. Cf. *djúprǫðull* in Bragi's verse quoted in *Gylfaginning*, ch. 1.
rǫf *f.* amber, in kennings for gold: *Rínar r.* 26/4, *spannar r.* 44/2 (subj. of *vensk*).
rǫnd *f.* shield-rim (or border of the shield-boss; by synecdoche = shield) 4/7, 16, 8/8, 9/5, 57/8; in kenning for sword, *rjóðvendill randa* 13/5, for shield (or description of the boards comprising the shields?) 59/4, for men 45/6, for battle 8/8 textual n. (see also explanatory n.). Cf. *Skáldskaparmál*, ch. 49.
rǫst *f.* (1) a distance of travel on land. (2) whirlpool; current 35/3 (obj. of *reist*); in kennings for ship, *rasta hlynr* (or *Huglar rasta hlynr*: *Huglar*

rǫst (1) = Hugl's [island's] way, i.e. sea) 19/2, *rasta hestr* 34/5, for ale, *r. jastar* 25/1.

sá (1) (søra) *sv.* with dat., strew 52/1 (see **niðr**); sow as corn 94/2.
sá (2) *pron.* demonstrative and anaphoric 1/6 (with *stillir*), 45/3 (with *fættir*, = Skúli); f. *sú* 13/7 (with *harða ferð*; if *harða* is not adv.); dat. *því* 14/7 (obj. of *fagna*); acc. *þann* that person 15/3.
saga *f.* story 35/13.
salr *m.* hall 90/1.
saltr *a.* salty 76/4 (with *stíg* = sea).
saltunna *f.* hall-vat, -barrel; in kenning for poetry 31/7 (*Hárs salr* = Valhǫll, *Hárs saltunna* = one of the containers of the mead of poetry [cf. *Skáldskaparmál*, *SnE* 1848–87, i. 216–24], its *hrannir* are the mead).
sama (ð) *wv.* impers., it befits 31/2 (with dat. of the person and inf.).
samhenda *f.* 'coincidental rhyme', rhyme falling on the same syllable as the alliteration (and producing identical syllables; cf. **liðhending**) 46/9.
samhending *f.* = **samhenda** 8/26, 47/9, 48/9.
samhendr *a.* having coincidental rhyme; n. as subst. as name of a verse-form 45/13.
samhljóðandi *m.* consonant 1/15, 32/11 (this is also the word used in *FGT* and *TGT*; = *málstafr* in *SGT*).
samkvæðr *a.* identically expressed, repeated, consonant 8/28.
samna (að) *wv.* gather together 17/19.
samr *a.* seemly, suitable, fine 38/6; with gen., suited to s-thing, subject, exposed to s-thing 72/4.
samstafa *f.* syllable 0/21, 28, 1/11, 30, 39 &c.; pl., (number of) syllables 9/10. This is the usual term in *FGT*, *SGT*, *TGT*.
samstǫfun *f.* syllable (= **samstafa**) 1/14, 19, 13/10, 29/10, 33/9, 34/10, 36/10, 65/11, 13 (*FGT* 224, 230, 240, *SGT* 63, n. to l. 42).
samþykki *f.* agreement, unity, peace (in kenning for gold, i.e. rings) 43/1 (cf. **søkk**).
sannan *f.* confirmation, proof, demonstration 18/10.
sannask (að) *wv. md.* prove true, become a reality 44/2.
sannkenna (nd) *wv.* use affirmatory or intensive attributives or adverbs 1/53; pp. n., as name of a verse-form 4/1 textual n.
sannkenning *f.* 'true description' 3/9, 4/9, 11, 12, 18, 21, 5/9, 10, 11, 6/20. (As here defined *sannkenning* refers to use of affirmatory, intensive, or evaluative epithets or adverbs, not to limiting or distinguishing epithets. The element *kenning* is used in the formal sense of the use of a description or compound containing two elements, whether metaphorical or not. Cf. *Skáldskaparmál*, ch. 67 (where the word refers to epithets describing innate or 'essential' qualities); *TGT* 100, 103 (probably an error for *mannkenning*), 108, and cf. pp. 320–1).

sannr *a.* true (literal, accurate) 4/9; n. as subst., truth, what is true 90/4.

sár (1) = *sá er* (rel.) 1/1 (either 'he who' as subj. of *lætr* or 'who' with antecedent *konungr*).

sár (2) *n.* wound 4/1, 10, 11, 94/5; in kenning for blood 60/2, for sword 61/2.

seðjask (saddask) *wv. md.* be sated, eat one's fill 9/4.

sefi *m.* thought, mind; in kenning for breast 6/8.

seggr *m.* man 4/2 (dat. of respect), 4/12, 5/2, 8/2, 18/7 (obj. of *grandar*), 8/20, 33/6 (dat. of respect), 72/3, 84/8 (acc. pl.), 97/7 (with *hverr*, subject of *muni*).

segja (sagða) *wv.* say, tell 42/13 (be exemplified), 85/5, 87/7.

segl *n.* sail 19/1.

seimgildir *m.* gold-payer (-giver or perhaps -increaser; cf. **gilda**), = Hákon 29/4.

seimr *m.* gold (thread) 47/1, 90/3, 97/8 (gen. with *gjǫflata*); in kenning for generous ruler (Skúli): *seima sneiðir* 71/2.

seimþverrir *m.* gold-diminisher, i.e. generous ruler, = Skúli 47/1.

seimǫrr *a.* gold-liberal 47/2.

seinn *a.* slow 6/22, 24, 17/18. Used of long vowels in *SGT* 66.

semja (samða) *wv.* put together, compose 68/3; arrange, don 36/3.

senda (nd) *wv.* send; i.e. give 91/6 (subject is *jarla beztr* = Skúli).

sendir *m.* sender, provider (impeller, user, wielder?); in kenning for war-leader (Hákon) 28/1.

senna *f.* quarrel; in kenning for battle, *sverða s.* 6/5.

serkr *m.* shirt; in kennings for coat of mail: *styrs s.* 7/8 (obj. of *rýðr*), *Skǫglar s.* 64/4 (in kenning for warrior).

setja (tt) *wv.* put 0/22, 1/10, 12, 18/19; set, place, position 1/37, 3/3, 7/11, 16/1, 46/10, 68/10; construct 6/12, 70/14; treat 84/7; *þar er fyrir sett* there is placed before it 54/11; *s. í* insert 33/10; *s. með* place with, next to, after, include with 49/10; *s. saman* make consistent 53/11, place together 57/10, compose 67/10; *s. út* extend 8/24.

setning *f.* arrangement, ordering, positioning 0/29, 1/25, 32 (?), 8/49 (?); prescription, rule 0/7, 8, 12, 16, 23, 1/24, 28, 44, 6/23, 8/45, 47; *til háttar setningar* which characterize the form, as the rule of the metre 62/10. Cf. *TGT* 36. In modern Icelandic the word means 'sentence'.

setr (rs) *n.* seat, abode; *buðlunga s.* = kingdom 15/8 (obj. of *stýra*).

sextánmæltr *a.* (*pp.*) containing sixteen utterances or sentences (cf. **mál (2)**) 8/51; n. as subst. 8/41 textual n. (apparently pl.), 9/1 textual n.

síðar *adv. comp.* later; *er ein er s.* last but one 1/39; *svá at tvær eru s.* before the last two 33/10.

síðari *a. comp.* second (of two) 1/38, 8/38, 48/10, 77/12, 81/9; succeeding, latter (sc. *vísa*) 16/10.

síðarstr *a. sup.* last 14/9, 15/10, 16/9, 24/9, 58/9, 70/13.

sigla (d) *wv.* sail; *vara siglt* it was not sailed, the sailing was not 27/8.

sig-Njǫrðr *m.* battle-Njǫrðr (or victory-Njǫrðr), kenning for warrior 55/8 (dat. with *varða*).
sík *n.* ditch, channel (= sea in kennings for gold) 17/1, 9, 27.
siklingr *m.* ruler; = Hákon 10/1, = Hákon and Skúli 67/3 (with *þá*-clause dependent on it), = Skúli 82/2, 90/1. Cf. *Skáldskaparmál*, ch. 64.
silfr *n.* silver; dat. in silver (utensils) 91/4.
sín *pron. refl. gen.*; *at sín* at his house 91/3.
sinn *poss. a.* his, its (own); *s. hljóðstafr* [there is] a separate (different) vowel [in each] 1/30, *í sínu vísuorði* in a different line 79/13.
sinni *n.* company 88/7.
sitja (sat) *sv.* sit 29/3, 88/7.
sízt *adv.* least of all, in no way 67/5; scarcely, not at all 84/6.
sjá (sá, pres. **sér)** *sv.* see 28/8, 58/5, 77/3, 93/8; subjunc. could see, could indicate 55/1; with acc. and inf. 56/7; *sjá til e-s* look for s-thing, expect s-thing 26/6.
sjaldan *adv.* seldom, not often 44/7.
sjaldnar *adv. comp.* more seldom, less often, fewer times 1/19.
sjár *m.* sea 17/17, 34/7; acc. over the sea (with *renna*) 101/7. Cf. **sær**.
skaði *m.* damage, destruction; *vinna gulli skaða* = distribute gold 47/8.
skafa (skóf) *sv.* scrape, smooth; pp. polished, burnished 8/4, planed 73/6.
skaka (skók) *sv.* shake 38/7; impers. with acc. (for pass.) 8/3, 9/5, 10/3 textual n., 19/1, 78/5.
skál *f.* bowl (as receptacle for ale or wine) 23/7, 91/5 (subject of *mætir*, i.e. it is held by).
skala = *skal* with neg. suffix, shall not, must not 68/2, 100/5.
skáldskapr *m.* poetry, the making of poetry; prosody 0/4.
skammr *a.* short 0/29, 70/11, 71/12; comp. *skemri* shorter 66/6 (predicative with *hal*), 72/10; less (than a full line) 8/29 (i.e. whether the repetition be of a full line or of something less).
skapa, skepja (skóp, pp. **skapat)** *sv.* and *wv.* create, cause (*e-t e-m* something for s-one) 19/6 (subject is *Rán*, obj. *deilu*, indirect obj. *flaustum*), 64/6; *s. e-t at e-u e-m* make s-thing into s-thing for s-one 33/2 (but possibly *at raunum* is an independent phrase ('in truth') and *skapat* has the sense 'caused, decided, fated').
skapt *n.* shaft (of spear) 52/2.
skarð *n.* cleft, gap 17/2 (*s. jarðar* = fjord(s) (collective), Firðir in Norway), 17/11, 12, 28.
skarpr *a.* sharp 32/1 (with *él*, or n. as adv. ?).
skati (pl. **skatnar**) *m.* man 7/2 (obj. of *spekr*), 18/3, 17, 37/7 (gen. pl. with *þengill*), 58/7 (with *vinr*); generous man, lord 71/7 (= Skúli); partitive gen. with *skýrstr* (perhaps parallel with *jarla* and to be taken with both *skýrstr* and *dýrstr*) 82/7, 94/7; *hringa s.* (i.e. the one generous with rings, = Skúli) 90/8.

skattr *m.* treasure; *Niflunga s.* = gold (see *Skáldskaparmál*, ch. 42) 41/8 (dat. with *hreytt*).

skaut *n.* sheet, skirt; expanse 95/8.

skefla (ð) *wv.* impers. with acc. s-thing is heaped up, made into heaps or drifts 76/2.

skeið *f.* warship, a large longship 21/5.

skelfr *a.* trembling 18/11 (cf. **skjálfa**).

skemma (ð) *wv.* shorten 8/19.

skenkja (nkt) *wv.* serve; pp. 91/4.

sker *n.* skerry, half-submerged rock in the sea; in kenning for gold: *liðar s.* (cf. *Skáldskaparmál*, ch. 32; part of kenning for ruler, *virðandi liðar skerja* = Skúli) 46/2; in kenning for teeth: *orða s.* 87/8.

skera (skar) *sv.* cut 4/3, 13, 27/1, 34/3, 59/1, 78/1; impers. (= pass.) 10/3.

skerðir *m.* diminisher, damager (he who makes a *skarð* in s-thing); in kennings for ruler (= Skúli): *hringa s.* i.e. generous prince 63/6 (subject of *rendi*), *Skǫglar serks s.* i.e. warrior 64/3, *Mistar lauka grundar s.* i.e. warrior 85/4 (dat. of advantage).

skíð *n.* long flat piece of wood; ski; in kenning for ships: *hlunna s.* 76/8 (dat. obj. of *hleypir*). Cf. **byrskíð**.

skilja (lð) *wv.* divide, part 17/18–19, 22/2; separate, differentiate 80/10; distinguish, understand, perceive 17/5, 31.

skip *n.* ship 19/2, 20/8 (acc. pl.[?]), 24/4, 38/5 (gen. pl. with *sýjur* or *þrǫmum*).

skipa (að) *wv.* with dat. arrange; *er e-u er skipat í* into which s-thing has been arranged or ordered 67/11; with acc., man, occupy 21/6; pres. p. *skipendr* those who man, crew (with gen.) 46/8; md., be occupied, get filled 23/6, 89/2.

skipreiða *f.* ship-levy estate, an estate providing the levy for equipping a ship 28/4 n. (obj. of *veitti*).

skipsbrot *n.* shipwreck 35/11.

skipta (pt) *wv.* with dat. (1) change, vary 1/48, 8/24, 23/14, 28/9, 30/9, 39/9, 80/10; impers. (= pass.) 31/9, 32/9, 34/9; *e-u er skipt* 8/50, 27/14 ('by which d. is varied'); md. be changed, varied 29/9 (cf. **breyta**; in *TGT* 111 *skipta* means exchange or transfer; cf. *FoGT* 120). (2) divide, share out 43/4 (cf. *TGT* 112); abs., deal, organize, arrange (obj. understood) 62/6 (see 62/5–8 n.).

skipun *f.* arrangement, ordering, classification; rule? (= *setning*?) or variation? 62/16. Cf. *TGT* 45, 76, 111, 112 (= *ordo*). See **skipa**.

skjaldborg *f.* shield-wall 16/4 (obj. of *setr*).

skjaldbrak *n.* shield-crash, i.e. battle; in description of ruler: *skjaldbraks mildr* battle-prodigal 28/3.

skjálfa (skalf) *sv.* tremble, shake 9/6, 18/4 (subject is *skjǫldr*; cf. **skelfr**), 18/17, 75/8 (subject is *ræði*).

skjálfhenda *f.* 'shivering', when the alliterating syllables in the odd lines are separated by only one other syllable 28/11 (obj. of *gera*), 14; *in forna s.* 34/12, *in nýja s.* 39/11.

skjálfhendr *a.* 'shivering', having alliterating syllables in the odd lines separated by only one other syllable 8/16, 35/13; n. as subst. 35/9.

skjóta (skaut) *sv.* with dat. shoot; put, cause to go 64/7; push (aside) 78/3; impers. (= pass.) spring, shoot, be shot up 13/8, there is shot 54/1; *ef henni er skotit í* if it (i.e. an occasional *minni alhenda*) is slipped into, if it is allowed to slip into 43/10.

skjótr *a.* fast 17/17–18; quick, i.e. unstressed or short (usually referring to resolution of stresses into two short syllables, or to enclitics and proclitics), 6/22, 7/11, 70/12 n., 83/10, 86/10 (cf. *SGT* 66); n. as adv. quickly 4/17, 17/31.

skjǫldr *m.* shield 7/6, 8/3, 9/5, 18/3, 17, 24/7, 61/4; *skjaldagi* = *skildir eigi* 54/2 (see Noreen 1923, § 396, Anm. 1; perhaps better to read *skjǫldungi*, as U, = *skjǫldum eigi*, instr.; see **halda**); in kennings for warrior 5/8, 30/2, for battle 55/4, 58/7.

skjǫldungr *m.* ruler; = Hákon 24/8 (subj. of *gefr*); = Skúli 70/8 (gen. pl. with *fremstr* 68/1), 91/8; *skjǫldungs niðr* = Skúli 52/1 (see **niðr**). (The word probably originally meant 'shield-warrior', but cf. *Skáldskaparmál*, ch. 64.)

skot *n.* shot, shooting 24/7.

skothenda *f.* half-rhyme, assonance 23/13, 42/10, 12, 43/10, 46/11, 52/9, 58/12, 63/11, 66/10, 75/10, 77/10.

skothending *f.* = **skothenda** 1/32–3, 41/11, 58/10, 68/13, 70/9, 76/11.

skothendr *a.* having half-rhyme 38/9, 51/13; n. as subst. 55/10.

skotna (að) *wv.* impers. *e-m skotnar e-t* s-thing falls to one's lot 44/1, one succeeds in s-thing, is fortunate in s-thing 61/8 (or *svá skotnar þat* as a separate clause, 'thus it turns out', see **svellir**).

skotskúr *f.* shower of missiles 16/3 (prepositional phrase with *setr*).

skrautfǫr *f.* splendid, glorious expedition 70/7 (obj. of *geta*).

skreyta (tt) *wv.* decorate 73/5.

skríða (skreið) *sv.* glide along (with acc. of the path) 6/2, 11, 13.

skriðr *m.* gliding movement 38/4 (gen. obj. of *biðja*), 44/18, 72/8 (obj. of *nýtr*).

skúr *f.* shower; dat. pl. (adv.) 62/7 ('in showers'; see n.); in kennings for battle: *darraðar s.* 52/1 (obj. of *sær*; if *darraðar* is taken with *skǫpt*, *skúr* here means 'shower of blows'), *stála s.* 55/5 (gen. with *Gauti*), *hjálma s.* 57/1, *Hlakkar s.* 64/3.

skutr *m.* stern 21/2.

ský *n.* cloud; in kenning for shield (part of kenning for battle): *lindar snarvinda s.* 32/2.

skyldask (ld) *wv. md.* pledge o-self, undertake (with *at* and inf.) 58/8.

skyli *m.* 'protector', king 2/4 (gen. sg. or pl.(?) with *ætt(stuðill)*; taken as proper name in *Skj* B ii. 61, cf. *Skáldskaparmál*, ch. 64), 28/3 (dat. with *þakkak*).

skylja (skulða) *wv.* wash (over) 60/1.

skýra (ð) *wv.* make clear; pp. *vann manndýrðir skýrðar* caused (his) virtues to be evident 44/4.

skýrr *a.* clear(-minded?), intelligent 70/7 (with *greppr*; indefinite); sup. 82/7 (either attributive with *Skúli* or predicative, parallel to *dýrstr*), 94/7.

skǫr *f.* hair, head 10/8.

skǫrungr *m.* outstanding one; = Hákon 2/8 (subj. of *felr* 2/5; or of *reiðir*, parallel to *deilir gulls*, if the subj. of *felr* is understood from the first half-verse), = Skúli 83/8 (obj. of *kalla*).

slétta (tt) *wv.* make level, make smooth or flat; pp. 17/2 (with *skarð*; *hafi* is instr.), 17/11, 28; slice off (the—metaphorical—obj. is buildings, which will be razed) 50/7.

sléttr *a.* flat, smooth 17/12.

slíðrbraut *f.* scabbard-path 6/2.

slíðrir *f. pl.* scabbard 6/9.

slíkr *a.* such; n. as subst., such an activity 27/4, 89/5, such activities 94/6, such a one (Skúli) 83/8; *slíkt er svá* so it is 82/1.

slíta (sleit) *sv.* tear apart 53/5.

slitna (að) *wv.* break, come apart 18/9; be broken 18/2, 15 (*dul slitnar at* i.e. it is revealed that, there is no disputing that, there are clear reports that).

slóð *f.* track, path; *s. geima*, i.e. the sea 71/1 (with *hleypa*: over the sea).

slyngva (slǫng) *sv.* throw; pp. *slunginn við* fastened, twisted together with or in 89/8.

smáorð *n.* (grammatical) particle 44/12.

smár *a.* small; n. as subst. *í smátt* into small pieces 89/7; comp. *smæri* lesser, less important or elaborate? or with shorter lines? 67/14.

snarla *adv.* swiftly, energetically 54/8.

snarr *a.* swift, keen, bold 4/8 (paraphrased *frækn* 4/17), 38/1, 43/2, 80/4 (with *fjǫrum*); of a poem, fine 92/2.

snarvindr *m.* keen wind; in kenning for battle, *lindar s.* (part of kenning for shield) 32/2.

sneiðir *m.* cutter; in kenning for ruler: *seima s.* i.e. distributor of gold, generous prince, = Skúli 71/1.

snekkja *f.* warship, 'snack' 20/6 (obj. of *rekr*), 38/2, 75/2 (obj. of *lætr*), 77/2, 98/4.

snerpa (t) *wv.* make sharp or harsh (*snarpr*); impers. (= pass.) become harsh 9/2.

snerpir *m.* sharpener; in kenning for war-leader (Skúli): *hlakkar s.* 42/4.

snerra *f.* onslaught, battle 53/6.

sníða (sneið) *sv.* cut (in two) 10/8.
snjallmæltr *a.* (*pp.*) clever in speech, eloquent 76/6.
snjallr *a.* brave or wise, clever 32/6, 42/5 (with *jǫfurr*), 67/3, 83/7.
snotr (trs) *a.* wise 41/2 (with *ýtum*).
snyðja (snudda) *wv.* hasten, go fast 77/1.
snýja *wv.* cause s-thing (*e-u*) to snow 62/1.
sog *n.* keel 22/2 (pl.: the keel parts of a ship?).
sókn *f.* advance, attack; battle; in kenning for swords 6/1, for missiles 62/2.
sóknharðr *a.* attack-, battle-hard 16/8 (with *vápnrjóðr*).
sóknvǫllr *m.* battle-field; in kenning for warrior (Skúli) or sword 61/7.
sól *f.* sun 17/21.
sólgit *see* **svelga.**
sólroð *n.* 'sun-reddening', dawn 77/1.
sómi *m.* honour 39/8 (with *allan*, obj. of *fær*).
sótt *pp.* of **sœkja.**
spakr *a.* wise 70/4 (with *mœti*).
spara (ð) *wv.* spare, hold back 80/3 (1st pers. pl., sc. Snorri), 92/4 (obj. *þat*, i.e. the praising; but see **þat**).
spekja (spakða) *wv.* make quiet 7/1.
spenna (t) *wv.* grasp, clasp 75/6.
spilla (t) *wv.* spoil 8/17; abs. be a blemish, be bad style 6/16 (opposed to *vel kveðnar*; equivalent of *nykrat*), 8/27, 58/16.
spjalli *m.* friend (one who converses with one [*e-s*]); *gumna s.* = Skúli 83/6.
spjót *n.* spear 16/6 (instr. with *stikar*), 33/7, 66/2 (instr. with *eyddi*).
spjǫll *n. pl.* tidings, news; i.e. accounts of Skúli's achievements 80/4 (obj. of *spǫrum*).
spjǫr *n.* spear 9/8, 10/2.
springa (sprakk) *sv.* burst, break; pp. *sprungit* broken, i.e. been distributed (by Skúli) 41/4.
sprund *n.* lady; *Hjaðninga s.* = Hildr, personification of battle 49/8 (subj. of *þiggr*).
spyrja (spurða) *wv.* hear, learn; with acc. and inf. (equivalent of 'that'-clause) 5/1, 47/3, 90/5; with acc. and a. or pp. (hear that s-one is s-thing) 5/7, 8/1, 61/7 (cf. **þrotna**); impers. (= pass.) *þat spyrr* this will be heard (reported) 89/4; pp. with *spjǫll* tidings (we have) heard 80/3; *s. at e-m* learn, hear about s-one 26/4 (i.e. they become famous).
spǫng *f.* metal plate or disc (part of mail-coat) 57/4.
spǫnn *f.* span; hand; in kennings for gold: *blik spannar* 40/8, *rǫf spannar* 44/2.
staðr *m.* place, position 1/30, 77/11 ('in each case', i.e. both in the odd and the even lines, with both kinds of rhyme); class, group, type, category 51/9, 58/17.

stafasetning *f.* spelling 0/28; arrangement of letters (i.e. alliteration) 1/9; (i.e. internal rhyme) 1/25, 27.

stafaskipti *n.* distribution, arrangement of staves (alliteration) 2/10, 9/10, 62/14, 63/11, 65/10, 67/9, 70/10, 74/10, 75/12, 79/10, 82/10, 88/10. Cf. **hendingaskipti**. (In *TGT* 65 the word means 'change of letters'.)

stafn *m.* stem (of ship) 73/6.

stafr *m.* (1) stave, staff; in kenning for warriors: *gunnveggs s.* 61/6 (dat. of respect with *leggi*). (2) letter, sound (at the end of a syllable or word) 1/31, 71/11, 73/10, 78/10; (at the beginning of a syllable) 41/9; pl. initial letters (or spelling overall?) 46/10; alliterating sound (stave) 1/10, 12, 13, 14, 16, 19; *at stǫfum* in alliteration 76/10, 11.

stál *n.* (1) steel; weapon 66/3; in kennings for warrior(s): *stála valdr* weapon-wielder (Skúli) 44/6 n., *stála steykkvilundar* 63/1; for battle: *stála skúr* 55/5. (2) inlay, inlaid (parenthetical) statement (cf. **stæltr**) 12/12 (cf. *TGT* 70). (3) stem, prow 20/1, 21/7, 75/4 (instr. with *hrinda*); in kenning for ship 31/1.

stálhrafn *m.* stem-, (prow-)raven (or -horse: Hrafn is the name of a legendary horse, see *Skáldskaparmál*, chs. 44 and 58), i.e. ship, in kenning for seafarer 59/5.

stálhreinn *m.* stem-, (prow-)reindeer, i.e. ship, in kenning for seafarer 28/7.

stallr *m.* perch; in kenning for hand: *gelmis s.* 2/7.

stamhendr *a.* stammering-rhymed 44/19, 45/10.

standa (stóð) *sv.* stand 13/4, 66/4; be placed 1/39, 45/11, 54/10, 58/14, 97/10; come 1/40, 32/9, 58/13, 65/11; appear, be 0/20; take place, continue 36/8; *s. af* arise from 37/8 (subj. is *þat* 37/7), 55/5; *s. fyrir* come at the beginning (of) 1/14, 19, 22, 41/9 (come in front of), 82/11 n. (come in place of, as?), come before 8/14; *s. í* be in the position of 44/12; *s. saman* be adjacent 23/10; *s. til* be capable of, stand in need of 16/16; md., come, be positioned 97/12; *standask nær* come, be positioned close to each other 28/10; *s. sem first (má)* come as far apart from each other as possible 28/12, 31/10.

stef *n.* refrain 70/12, 15, 81/5 (obj. of *stæra*; referring to verse in general?—but cf. Fidjestøl 1982, 248, where it is suggested that the reference is to the *stef*-like repetition in 82/7–8 and 94/7–8).

stefjamél *n.* refrain passage, passage enclosed by a refrain 70/14, 15 (gen. of respect).

steflauss *a.* lacking a *stef* or refrain 35/13.

stefnir *m.* director, steerer; in kenning for ruler (Skúli): *stálhrafna s.*, i.e. seafarer 59/5.

steindr *a.* (*pp.*) coloured 10/3.

steinn *m.* stone; instr., by stone 102/7.

sterkr *a.* strong 4/2, 12 (complement), 7/8.

steykkvilundr *m.* in kenning for warriors, *stála s.* steel-impelling (throwing) trees 63/1 (dat. with *kendi*). Cf. **støkkvir**.

stígr *m.* path; pl. in kennings for breast: *sefa s.* 6/8, *hugar s.* 6/13; in kenning for sea: *kjalar s.* (dat. obj. of *kasta*) 76/3.

stika (að) *wv.* fence (make a fence round with palings), enclose in a fence 16/5.

stikkalag *n.* 'needle-metre'?—probably an error for *Starkaðar lag* 97/11, but see note and cf. *LP* s.v. *stikki*; *NN* 2095; also the poem-names *Sǫrlastikki, Haraldsstikki* (*Flb.* i. 307, *Hkr* iii. 181).

stiklir *m.* thrower (*stikla*: cause to jump or fly); in kenning for (generous) ruler (Skúli): *mens s.* 60/5. Cf. **menstiklir, hoddstiklandi**.

stillir *m.* ruler; = Hákon 1/6, 7/7, *hersa s.* 29/7 (dat. with *hollr*); = Skúli 60/7, 66/3 (gen. with *dólgum*), 76/6, *s. Mœra* 81/6 (dat. of advantage, for); collective, = Hákon and Skúli 102/8; in kenning for warrior, controller, operator (of swords, = Hákon) 2/1.

stinga (stakk) *sv.* with dat. thrust; *s. brott* dispose of, give away 47/4 textual n. See **þinga**.

stingr *m.* prow, in kenning for ship: *stinga hjǫrtr* 73/7 (the pl. may refer to spikes fitted on the prows of ships).

stinngeðr *a.* firm-minded 31/2 (with *þengils*).

stinnr *a.* stiff, unbending 21/8, 60/6 (n. pl. with *fólk*; or emend to *stinnr*, with *stiklir*, 'unyielding'; see 60/8 n.); severe, sharp, painful 4/1, 10 (complement).

stjóri *m.* controller; in kenning for ruler (Skúli): *s. dunu geira* 53/1.

stóð *n.* stallion (accompanied by mares); in kenning for ships: *s. Róða* 21/4 (also by word-play p. of *standa* sv., 'stand still', in contrast to *hljóp*).

stórgjǫf *f.* great gift 95/4 (obj. of *munða*).

stórlæti *f.* munificence 28/8.

stórr *a.* great 4/10, 53/2; a great deal of 11/6; n. as adv. *stórt* mightily 15/7 (or n. as subst., 'he undertakes great things'?); *stórum* greatly 4/1, 11; comp. *stœrri* greater 40/3.

straumr *m.* current (of sea) 101/6 (obj. of *reist*).

strjúka (strauk) *sv.* stroke 22/1.

strúgr *m.* dignity, pride; or indignation, anger?—in kenning for wine, *strúgs galli* 25/7.

strýkva *sv.* stroke 76/1.

strǫnd *f.* shore; in kenning for shield: *brands s.* 59/4.

stuðill *m.* (1) prop, support (i.e. the alliterating staves in the odd lines of verses) 1/14, 15, 21, 28/11 (nom. pl. with *þeir*), 31/10, 97/10, 11, 12. Cf. **hljóðfyllandi, hljóðfylling**. (2) 'buttress', an additional line at the end of a stanza (?) 8/32 n.

stuðning *f.* support (i.e. intensive prefix) 4/22, 5/9, 11.

stúfhendr *a.* 'stump-rhymed', having the (first) rhyme-word docked (shortened by a syllable, i.e. monosyllabic); n. as subst. 73/11.

stúfr *m.* stump, something docked; catalectic (verse) 48/11; *meiri s.* 49/10, *hinn mesti s.* 50/9. Cf. **stýfa**.

stundum *adv.* sometimes 1/40, 41, 27/15.

styðja (studda) *wv.* support; pp. 'supported as it is' 102/7 (with *fold*; *steini* is instr.); use a support (*stuðning*), i.e. an intensive prefix with an adjective or adverb 1/53 (absol.), strengthen (with an a. or adv.) 4/9.

stýfa (ð) *wv.* 'make into a stump', dock, apocopate; pp. *stýfðr* catalectic 49/9, 50/9, 10, 51/10, 12; of a word rather than a line, shortened by one syllable, monosyllabic 75/11, 76/11, 77/12; shortened, abbreviated 81/10, 93/9.

stýra (ð) *wv.* with dat. govern, rule over 14/6, 15/6 (obj. is *grýttu setri*); steer 63/3.

stýri *n.* steering-oar, rudder 35/4 (instr.), 74/4.

stýrir *m.* steerer, controller, user; in kennings for seafarer: *hranna hádýra s.*, i.e. viking warrior, man 8/37, *stálhreins s.*, i.e. Snorri 28/6 (dat. with *gjǫf*); for war-leader (Skúli): *hjaldrs s.* 39/6 (dat. with *gaf*).

styrjǫkull *m.* battle-glacier, -icicle, i.e. sword 60/5 (obj. of *venja*).

styrkr *a.* strong, powerful 52/7 (with *mót*, probably attributive not predicative: 'powerful meetings of metal take place under the standards').

styrr *m.* tumult, battle 10/3, 55/6; warfare 62/8; in kennings for ruler: *styrjar deilir*, = Hákon 21/7, 28/7, *styrjar valdi*, = Skúli 63/2, *styrs stœrir*, = Skúli 68/6; in kenning for battle: *hjálma hyrjar s.* 58/2 (gen. with *til*); for swords 85/7; for coats of mail, *styrs serkir* 7/7.

styrvindr *m.* battle-wind, i.e. warfare 59/6.

stæltr *a.* (*pp.*) inlaid, intercalated 11/10; n. as subst. 12/1 textual n. Cf. **stál**; *stæla* means 'to hammer steel into'. Cf. *TGT* 70, 113, *FoGT* 136–7.

stœra (ð) *wv.* make great, increase; carry out (deeds) 53/1; cause 63/7; create, perform (poetry) 31/8, 81/5. Cf. **auka, efla, gilda, herða, remma**.

stœrir *m.* increaser, creator, performer; in kenning for war-leader (Skúli): *styrs s.* 68/5 (dat. of advantage). Cf. Latin *auctor* (from *augeo*).

stǫðva (að) *wv.* cause to stop, put a stop to 17/7 (with person being restrained in dat.), 17/16, 23, 32.

stǫng *f.* pole (for standard or banner) 33/3, 52/6 (collective).

støkkva (stǫkk) *sv.* fly, spring 10/4; of a sword causing a wound 50/3.

støkkvi-Móði *m.* in kenning for ruler (Skúli): *styrjar glóða s.* the Móði who makes fly or impels (i.e. wields) or scatters (i.e. gives) swords 85/8 (gen. with *dýrð*). Cf. **steykkvilundr**.

støkkvir *m.* impeller; in kenning for seafarer (Snorri): *stáls dynblakka s.* 31/1 (dat. with *samir*).

støkr *m.* flight 35/5 (subject of *óx*).

súð *f.* planking (on a ship's side) 27/2.
sumar *n.* summer 23/2.
sumr *a. pron.* some; n. as subst. some of this, 'this is in some ways' 65/14.
sund *n.* sound, sea; in kenning for blood: *s. unda* 60/8 (obj. of *lætr* or *fylla*, see n.).
sundr *adv.* apart 10/2, 89/7.
sundra (að) *wv.* separate, tear apart; md. for pass. 9/7.
sundrgreiniligr *a.* containing diversity of meaning, antithetical 17/26.
svá *adv.* thus; *nema s. at* if it were not that (or unless in this way that) 56/3; *þar svá at* = *svá at þar*, or as rel. = where 58/3.
svalr *a.* cool 18/8, 12, 19, 35/4, 44/8 (with *branda*, a metaphor for gold (ornaments); an example of *refhvarf*, cf. 18/8), 90/3 (again referring to gold).
svanfjall *n.* swan's mountain, i.e. wave; in kenning for ships 83/4.
svanr *m.* swan; in kenning for waves 76/5.
svartr *a.* black, of ships 73/8, 76/5 (with *skíðum*).
sveimþreytir *m.* turmoil-labourer, -performer, i.e. warrior (= Skúli) 32/5 (reading *seim-*, as TWU, the term would mean 'gold-spender', i.e. generous ruler).
sveit *f.* troop, company (of men) 44/15, 83/7, 91/3.
sveiti *m.* sweat; in kenning for blood: *sæfis s.* 54/5; = blood 6/6, 32/5 (dat. with *bjó*).
svelga (svalg, pp. **sólgit)** *sv.* swallow 51/7.
svelja (svalða) *wv.* be(come) cold 35/7.
svell *n.* (lump of) ice; in kenning for sword, *sóknvallar s.* ?61/7 (instr. with *þrotna*). See next.
svellir *m.* heaper up, causer of swelling, in kenning for warrior: *sóknvallar s.*, i.e. he who heaps up corpses on the battlefield, or he who increases battle (cf. **ógnsvellir**); = Skúli 61/7 (dat. with *skotnar*, or dat. of agent ('at the hands of, before') with *þrotna*). Cf. *NN* 175, 1319 and see **svell.**
sverð *n.* sword 2/13, 4/5, 15, 6/9, 15 (subj.), 8/4, 10/1 (instr.), 17/30, 33/2, 57/2 (instr.), 58/4, 63/7 (instr.), 65/7; in kennings for battle 6/5, 16/7, for tongue, *góma s.* 85/3 (instr.).
sverðtog *n.* sword-drawing, i.e. battle 54/6.
sviðr *a.* wise 6/1.
svífa (sveif) *sv.* glide 20/2 (impers. with dat.?); glide round (from hand to hand, *at gjǫf* as a gift, freely) 23/8; glide through (the path in acc.) 50/5.
svipa (að) *wv.* with dat. cause to vibrate or flex, jerk; or intrans., sweep (forward) 38/5.
svǫrðr *m.* (1) thong, rope made of hide 21/3 (obj. of *herða*). (2) scalp; in kenning for head: *svarðar land* 57/6 (obj. of *hnykkja*).
sýja *f.* rivetting, clinching, line of rails (of the side of a ship) 38/5.

sýna (d) *wv.* show, demonstrate, exemplify 4/18, 7/11, 8/10, 16/15, 62/16; display 6/20; 17/25; md., be apparent 16/16.
syngva (sǫng, sungu) *sv.* sing; of noise of weapons 66/3.
sýnn *a.* obvious, clear to see, unequivocal; comp. 40/7.
sæfuni *m.* sea-blaze, i.e. gold 46/6 (instr. with *fremr*).
sæll *a.* happy, blessed 8/34.
sær *m.* sea 13/4, 17/13, 21; in kenning for gold 17/6, 20, 32. Cf. **sjár**.
sætt *f.* settlement, agreement 17/20.
sœfir *m.* queller, killer; = sword in kenning for blood: *sœfis sveiti* 54/5.
sœkir *m.* attacker, enemy; in kenning for (generous) ruler (Hákon): *s. síks glóðar*, i.e. enemy of gold, generous giver 17/1, 27; by word-play also 3rd pers. sg. of *sœkja* wv., 'attacks' 17/10.
sœkja (sótta) *wv.* seek 101/2, with suffixed pron. *-k* 101/1, 3; *s. heim* visit (s-one in his home) 13/6, *láta e-n sótt heim* pay s-one a visit 93/6; attack 17/11; *s. fram* advance, engage 64/2.
sœmð *f.* honour 27/6, 90/2 (*til* in).
sǫgn *f.* crew 24/4 (obj. of *gleðr*); *bragna s.* troop of men 59/8 (gen. pl. with *oddviti*).
sǫngr *m.* song; in kenning for battle: *at sverða sǫngvi* 16/7.
sǫnnunarorð *n.* confirmatory or corroboratory word, word of emphasis; intensive 5/11 (cf. **sannan, sannask**).
søkk *n.* gold, treasure ?43/1 (instr.; *samþykkjar* descriptive gen. (peace-bringing), cf. *friðbygg*. See *NN* 2179). See **søkkvir**.
søkkva (sǫkk) *sv.* sink 21/8.
søkkvir *m.* destroyer, in kenning for gold (objects): *samþykkjar s.* 43/1 (instr.). Cf. **søkk**.

taka (tók) *sv.* take 23/4; seize 18/11; receive, get 45/3; with inf. begin to 18/4 (subject is *skatnar*), 17, 38/3, go and 49/1; with *at* and inf. 88/3; *t. af* shorten, decrease 27/16, omit 49/9, 62/10, 65/13, base on, derive from 86/9, 88/9, 90/9, 92/9; *t. með* include, be consonant with, link with 48/9; *t. til máls* use for the sense 18/14; *t. ór* subtract, leave out 34/11, 42/9, 65/11 (with dat., from s-thing), take away from, subtract from 8/19, 62/11; *t. upp* take up, begin, raise (a topic) 6/14; *t. upp* (*vísu*) interpret (a verse), read in order 17/27.
tala (1) *f.* number, enumeration 0/15, 16, 19, 20, 21, 6/23, 8/46, 47, 48. Cf. *TGT* 50, 74, 76, 77, 85, *FoGT* 135, 142, where the word refers to grammatical number.
tala (2) *f.* speech; in kenning for tongue 81/4.
tala (3) (að) *wv.* speak 90/4.
téa (ð) *wv.* periphrastic aux. with inf.; *tér bergja ok rjóða* does taste and redden 11/7.
teitr *a.* happy 12/4 (with *Hákun*), 11.

telja (talða) *wv.* declare; with double acc., declare s-thing to be s-thing 89/5; count, enumerate 100/3.

tíð *f.* time; tense 8/25 (cf. *TGT* 76); (metrical) quantity? 16/13 n., 23/11 n. Cf. *TGT* 52–3, *GT Prologue* 154.

tíðaskipti *n.* variation of tense 8/41. Cf. *TGT* 77.

tíðr *a.* frequent; *tíð erumk* (s-thing) is frequent with me, i.e. I often do it, I am good at it, I like doing it 8/36; dat. pl. as adv., frequently, repeatedly, constantly 44/16; sup., most common, usual 70/15; sup. n. as adv. generally 65/9, 83/9.

tiggi *m.* prince, ruler (= Hákon) 12/3 (dat.); *tiggja sonr* = Hákon 18/7, 20; = Skúli, *t. grundar* 52/4 (dat.), 62/1, 74/2 (gen. with *hlunnvigg*), 90/3.

til *prep.* with gen., to 27/3, 88/1, 90/1; towards 63/3; for 26/6, (purpose) 27/8, 58/3 (to bring about), 88/5, as 53/7, 70/12, of 58/1 (with *væni*); in (as a result of) 90/2; about 100/2; (time) until 30/7.

tilsagðr *a.* (*pp.*) 'annotated'; n. as name of a verse-form 24/10, 25/1 textual n. (kennings in the verse are explained in the following line).

tiltekinn *a.* (*pp.*) 'linked'; n. as name of verse-form 15/1 textual n., 38/11.

tími *m.* time; in kenning for battle (time of swords), *sára ára t.* 61/1.

tírr *m.* glory 12/3.

tírœðr *a.* counted in tens (i.e. a decimal rather than duodecimal hundred) 100/3.

tjald *n.* tent, awning, curtain; in kenning for shield, *Hǫgna meyjar* (Hildr's, battle's) *t.* 49/4; for breast, *móðsefa t.* 50/2.

toginn *a.* drawn (of a sword) 58/4.

tómr *a.* empty 23/8 (with *hǫll*). *NN* 1305 takes *tóm* here as the noun *tóm* n., 'leisure', used adverbially with *svífr* = during the period of leisure.

traustr *a.* reliable 4/3, 13.

treystir *m.* tester (truster? encourager?), in kenning for ruler (= Skúli): *fólka t.* 34/1 (cf. *reynir*, *Skáldskaparmál*, chs. 31, 47 and *LP*, s.v.).

trollsháttr *m.* 'troll's verse-form' 62/17.

trúa (ð) *wv.* with dat., trust 26/5.

tryggva (ggð) *wv.* secure, entrust (*e-t e-m* s-thing to s-one) 43/6.

tunga *f.* tongue 81/3.

tvennr *a.* double, having two divisions 0/9, etc.; twofold 53/9; pl., having two sets (pairs) 17/35, 19/10, 44/9, 60/9.

tvíkendr *a.* (*pp.*) doubly modified, with double (two) determinants 2/11, 12.

tvíklypt *pp.* (cf. *klippa* wv. 'clip', see Noreen 1923, § 85) repeated 45/11. U has *tvíkylft* (written *-kylpt*), and this word is perhaps related to *kylfa* f., 'club'; the original may have had *tvíklifat* (cf. **klifat**). *Tvíklippa* and *tvíklifa* appear as variants with the meaning 'repeat' in *Konungs skuggsjá* (1920), 121/9.

tvíkveðinn *pp.* repeated; *t. at* repeated as to, (there is) repetition of 45/9.

tvíriðinn *a.* (*pp.*) 'double-twisted', doubly strengthened, having two qualifiers (adverbs) to the adjective or adverb; n. as subst. 4/22.

tvískelfðr *a.* (*pp.*) 'double-shaken', having alliterating syllables close together and a heavy syllable between them twice in the half-stanza; n. as subst. 27/16, 28/1 textual n. The word is used in Hallar-Steinn's *Rekstefja* 35 (*Skj* A i. 552).

tvístýfðr *a.* (*pp.*) 'doubly docked', catalectic in two lines 51/11.

tysvar *adv.* twice 1/14, 47/9.

tøgdrápuháttr *m.* 'journey-poem form' 70/12. The quality of the first vowel is uncertain; cf. *tog* n., 'rope', *tøgr* m., 'ten' (of a poem of ten stanzas?), German *Zug*, 'journey'. In *ÍF* xxvii. 308 the poem *Tøgdrápa* (by Þórarinn loftunga; probably the first poem to use this form) is connected with King Knútr's journeying in Norway. Sighvatr's *Knútsdrápa*, the next poem in the form, is also a journey-poem.

tøgdrápulag *n.* 'journey-poem metre' 67/16 (see **tøgdrápuháttr**, which apparently means the form of the poem as a whole, while *tøgdrápulag* and *tøglag* refer to the form of individual verses, see JH–AH 60).

tøglag *n.* 'journey-metre' (evidently = **tøgdrápulag**) 68/14, 70/11, 86/9.

tøgmæltr *a.* (*pp.*) 'journey-spoken', in *tøglag*; n. as subst. 70/10.

tøgr *m.* ten; *sex t.* with partitive gen. = 60, 67/4.

tœja (tœða) *wv.* help; abs. 81/3.

uðr *f.* = **unnr** wave 22/7.

úlfr *m.* wolf (i.e. Fenrir, in kenning for Óðinn) 3/1; dat. pl., for wolves 11/1.

umgjǫrð *f.* fittings, scabbard 6/10.

una (ð) *wv.* with dat., rejoice in 4/8, 17, 11/4; *vel una sínu* be content with one's lot 8/37; with inf. be happy to do s-thing 72/6.

und (1) *f.* wound 10/2, 56/5 (pl., obj. of *gefr*); in kenning for blood: *unda sund* 60/8.

und (2) *prep.* with acc. and dat., under 21/4, 32/7, 38/8, 49/3 (with *tjald*), 52/7, 53/4, 64/7 (with *il*), 72/5, 95/8.

undarliga *adv.* amazingly 4/17.

undgagl *n.* wound-gosling, i.e. raven 62/2 ('the bird of prey is aware of this', i.e. that corpses will be available as a result of the chieftain's warfare).

undinn *see* **vinda**.

undrum *adv.* amazingly (= *undarliga* 4/17) 4/7.

ungr *a.* young; of Hákon 1/6; with *herstefnir* (i.e. Skúli) 51/2, with *Skúli* (68/1) 70/8 ('when still young'); of Skúli 77/3, of Hákon and Skúli 98/7.

unna (ann, unna) *pret.-pres. vb.* love 19/11; *u. e-s* grant s-thing, be pleased for s-thing to be so 82/3.

unnr *f.* = **uðr** wave, in kennings for ship: *unna Gyllir* 19/4 (also by word-play *unna* 'love' 19/11), *unna dýr* (pl.) 28/5.

upphaf *n.* beginning 1/40, 28/11, 37/10; opening 70/15; i.e. foundation (?) 1/43.
upphafsstafr *m.* initial consonant or sound 1/31, 36, 41/12.
úthlaupsmaðr *m.* highwayman, robber 66/1.
útsker *n.* outlying skerry, offshore rock 35/11.
útstrǫnd *f.* outlying coast 79/2.

váð *f.* cloth; sail 20/5, 77/7 (obj. of *greiða*).
vaða (óð) *sv.* advance 33/3, 52/6 (*at*: to); step, pace (see **veðr**); rush forward 86/6 (*at gjǫf*: to be given, i.e. are quickly distributed); *v. fram*, of a poem: be presented 69/6.
váði *m.* causer of danger or harm; enemy; in kenning for Þórr: *orms v.* 3/5, for Óðinn: *vitnis v.* 8/36. Cf. *Gylfaginning*, ch. 51.
vaka (ð) *wv.* be awake 38/3.
val *n.* choice(st) 25/7 (with gen. pl.).
valbjórr *m.* slaughter-, carnage-beer, i.e. blood 11/6.
valbroddr *m.* slaughter-point, spear; gen. pl. with *ern*, eager to use spears, skilful with spears; or perhaps parallel to *hrænaðra*, gen. with *oddum* 79/6 n.
vald *n.* power; use of or desire for power, arrogance 37/2 (obj. of *banna*).
valda (olla) *sv.* with dat. cause, bring about 40/5; *v. e-m e-u* cause s-thing to s-one 5/3; rule over, possess 12/1, 10 (objects are *heiti ok hǫlðum*); pres. p. *valdandi* ruler: *hers v.* = Skúli 49/4.
valdi *m.* ruler, wielder; in kennings for rulers, *vígfoldar vandar v.* sword-wielder, = Hákon 30/5 (vocative); *styrjar v.* battle-leader, = Skúli 63/2.
valdr *m.* wielder; in kennings for warrior/ruler: *v. skjaldar* = Hákon 5/8, *v. stála* = Skúli 44/7 (or this could be the nom. **valdi** here, see n.).
válka (að) *wv.* handle; roll in the mind, ponder, plan 17/5 (pp. with *friðlæ*; *fljótt válkat* = rash), 17/18 (in the sense 'ponder, weigh', or (?) 'hover'), 17/30.
valland *n.* falcon-land, resting place of falcon, i.e. arm; in kenning for gold, *vallands brandr* 44/8.
valr (1) *m.* the slain, the fallen 6/7, 66/4.
valr (2) *m.* falcon; in kennings for arms: *vals reitir* 42/3, *vala leiðar* 48/8.
valstaðr *m.* falcon-perch (-stead), arm; in kenning for gold 86/7.
ván *f.* hope 33/6 (dat. obj. of *hrauð*).
vandaðr *a.* (*pp.*) finely (painstakingly) wrought, elaborate, done with care 1/42.
vandbaugr *m.* 'wand-ring'; *vandbaugs sendir* = *baugs vandsendir*, sender or provider or wielder of the wand of the shield (*baugr* [part of a shield, the boss or its surround] by synecdoche = shield, its wand is a sword), i.e. the king 28/1.

vandbaugskaði *m.* 'wand-ring-damager' = *baugs vandskaði*, damager of the wand of the shield, sword-damager, warrior 86/8 (pl., obj. of *drekkr*).
vandliga *adv.* precisely, completely 4/19; carefully 16/16.
vandr *a.* difficult 16/13; sup. most demanding, choicest 44/10 (cf. **vandaðr**).
vápn *n.* weapon 17/29, 96/4 (pl.); in kenning for battle 65/1.
vápnrjóðr *m.* weapon-reddener (i.e. with blood), warrior, = Hákon 16/5.
vara (1) (ð) *wv.* impers. *e-n varir e-s* s-one expects s-thing 92/3.
vara (2) (að) *wv.* warn 20/8 (word-play; cf. **vǫr**); md. be on one's guard against, avoid (*þat*: this practice) 44/13.
vara (3) = *var* with neg. suffix 27/8.
varða (að) *wv.* defend 18/4 (obj. is *foldir*), 18/11, 17; *v. e-t e-m* defend s-thing against s-one 55/8.
vargr *m.* wolf 11/7, 18/15, 51/5, 56/4 (dat. of respect or advantage), 96/3.
varmr *a.* warm 6/7.
varrsími *m.* (or **-síma** *n.*) thread or line of wake 35/2 (acc.).
vás *n.* hardship, wet and cold 72/3 (gen. with *samir*).
vatn *n.* water 6/11, 17/10.
vátta (að) *wv.* with dat., bear witness to, report 27/6.
vaxa (óx) *sv.* grow, increase 9/1, 35/5, 61/2; grow in power, gain 26/8.
vé *n.* standard, banner 52/5 (subject of *geisa*).
veðr *n.* weather, wind 20/7 (also by word-play 3rd pers. sg. of *vaða* sv., 'pace, step deliberately'), 35/12; in kenning for battle: *Skǫglar v.* 54/1.
vefja (vafða) *wv.* fold, wind; hesitate 64/1 (with *at* and inf.).
vefr *m.* cloth; sail 20/6 (also by word-play 3rd pers. sg. of *vefja* wv., 'fold').
vega (vá) *sv.* fight 8/23.
veggjaðr *a.* (*pp.*) 'wedged'; n. as subst. 32/12, 33/1 textual n.
veggr *m.* wall; sail 78/6 (obj. of *skekr*); in kenning for shield: *Sigars v.* 59/2 (pl. obj. of *skerr*).
veghrœsinn *a.* 'glory-boasting', proud of one's glory 5/7.
vegr (1) *m.* honour 67/5 (dat. obj. of *orpit*).
vegr (2) *m.* way 1/45, 49, 6/15, 8/39, 9/11; in kenning for sea: *Haka v.* 76/7.
veig *f.* drink, a filled cup 25/7.
veita (tt) *wv.* give 25/6 (obj. is *heilsu*), 28/1, 90/3, 94/5.
veizla *f.* feast 88/6.
vel *adv.* well; with sup. easily, by far 93/3.
vél *f.* trick, deceit 17/20.
véla (t) *wv.* cheat, use deceit on; entrap 62/4; with gold as obj.: cause to be lost, i.e. give away generously 40/8.
veldi *n.* power 14/4.
velja (valða) *wv.* choose, select, arrange (*saman* together) 16/12; select (*e-t e-m* s-thing for someone [as a gift]) 41/1.
vell *n.* gold; pl., gold ornaments 46/3 (instr. with *gleðr*), 67/5, 98/2, 99/5.

vellbrjótr *m.* gold-breaker, i.e. generous lord, = Hákon 16/6 (subj. of *þryngr*).

vellbrjóti (**-broti** (so TWU) would be more normal) *m.* gold-breaker, i.e. generous lord, = Skúli 46/3. Cf. **gullbroti**.

velta (lt) *wv.* with dat. cause to roll 65/1.

venja (1) *f.* custom 26/2.

venja (2) (vanða) *wv.* accustom (*e-t e-u* s-thing to s-thing) 48/7 (i.e. gives frequently), 60/6 ('accustoms swords to blood, fights frequently'); md., become customary, become normal or common 44/3; with *at* and inf., accustom o-self to s-thing, do s-thing frequently 53/1.

ver *n.* sea 82/6.

vera (var) *sv.* be 13/7; *er* = there is 58/1, it is 83/5; *sem er* as it is 87/7; with pp. 15/2, = has been 80/1, 97/3, 100/3, with neg. suffix *vara* (impers.) 27/8, of customary action 50/7; impers. referring to customary action 41/7, 48/3; *eru* = take place 52/7; with suffixed pron. *emk* 27/4, *vartu* 30/7; *ró* = *eru* 99/1.

verbál *n.* 'sea- pyre' i.e. fire of sea, kenning for gold 46/2 (obj. of *gefr*), 93/4 (obj. of *gefr*).

verða (varð) *sv.* become 68/1; come to be 41/12, 77/10; turn out to be 65/14; with pp. (forming pass.) 32/3, 33/1; with inf., have to 32/7.

verðr *a.* worthy 37/3 (with *ferð*; but according to *NN* 1311, this is *verð* n. repayment, reward, obj. of *fest*); with gen. (*bana*) deserving of 65/4 (with *klett*); comp. with gen. (*hæra hróðrar*) 98/1 (with *unga jǫfra*).

verðung *f.* court, king's or earl's following 46/4.

verja (1) (varða) *wv.* defend 3/1, 10/1, 14/3, 17/1, 10, 11, 28; with suffixed pron. *varðak* 8/22 n.

verja (2) (varða) *wv.* enclose, clothe, cover; of covering men's arms with gold rings, *v. e-m arm* 42/7, *v. e-t auði* 48/1.

verk *n.* deed 36/4 (pl., subject of *frǫmðusk*), 53/2.

verki *m.* (literary) work, composition (i.e. a poem, poetry) 8/31, 58/14.

verpa (varp) *sv.* with dat. throw 42/4; impers. with dat. for pass. 9/2; pp. *orpit* 67/7.

verstr *a. sup.* worst; *vellum v.*, i.e. most hostile or harsh to gold, most eager to be generous with it 99/5.

vestan *adv.* from the west (or in the west?) 35/1.

véstǫng *f.* banner-pole, standard 36/2 (subject of *reistisk*).

vetr *m.* winter 23/2; acc. of time 84/8.

við *prep.* (1) with dat. against 16/3 (prep. phrase with *setr*); at, in the face of 88/4; with, i.e. where he is 44/7 (but according to *NN* 3146, postposition with *herfjǫlð* (acc.); see n.). (2) with acc. against 34/7, 87/8; across, over 23/3 (dependent on *gǫngu*); together with (or by means of; with *slungit*) 89/8; because of 45/5.

víða *adv.* widely, over great distances 16/5; widely found, in many places 58/14.

víðir *m.* 'expanse', sea 74/6 (obj. of *brjóta*).

viðr (1) = *vinnr*, see **vinna**.

viðr (2) *m.* tree; by synecdoche = ship (gen. sg. with *skriðar*) 38/4, 72/8 (cf. **eik**); in kennings for warriors, *v. fólkhǫmlu* 24/6, *v. randa* 45/6 (dat. of respect with *armr*, which is collective).

viðrhending *f.* accessory rhyme (the second of two internal rhymes in a line) 1/38.

víg *n.* battle 41/7.

vígdjarfr *a.* battle-bold, battle-daring; sup. 99/6.

vígdrótt *f.* war-band 16/2.

vígfold *f.* battle-land, i.e. shield 30/5.

víggjǫll *f.* river of battle, i.e. flowing blood 6/8 (Gjǫll is the name of a mythical river in *Gylfaginning*, chs. 4 and 49).

vígrakkr *a.* battle-bold 28/2.

vígrœkjandi *m.* cultivator of battle, = Skúli 64/2.

vígsárr *a.* battle-wounded (of the breast) 51/6.

vili *m.* will; in kenning for breast 51/5.

vilja (ld) *wv.* aux. with inf. want to; with suffixed pron. *vilk* 85/1; 3rd pers. sg. *vill* tries to 22/3, *vill svá* wishes it so 51/8; with *at* and subjunc. 41/5.

vín *n.* wine 25/8, 91/4 (obj. of *drekka*).

vinda (vatt, pp. **undinn)** *sv.* wind, twist; pp. twisted 45/1 (with *brot*).

vindr *m.* wind 20/5 (also by word-play = 3rd pers. sg. of *vinda* sv. 'twist').

víngerð *f.* wine-making, wine-product; *vitnis váða v.* = Óðinn's wine-making, i.e. the making of (the mead of) poetry 8/37 (see *Skáldskaparmál*, *SnE* 1848–87, i. 222–4).

vinna (vann; 3rd sg. pres. **viðr)** *sv.* do; *v. e-m e-t* cause s-one s-thing 17/25, 19/5, 47/7; with pp., cause s-thing to be done 37/1 (*vann gjald* [or *verð*, see **verðr**] *fest*), cause s-thing to be made s-thing 44/4 (*vann manndýrðir skýrðar*).

vinr *m.* friend; in kenning for Óðinn: *Míms v.* 3/4; for King Hákon: *gotna v.* 11/5, *drengja v.* 14/2; for Skúli: *aldar v.* 54/7, *skatna v.* 58/7, *bragna v.* 90/6.

virða (rð) *wv.* evaluate, consider (s-thing to be s-thing, *svá*) 25/2; *v. e-n e-s* consider s-one worthy of s-thing, honour s-one with s-thing: *er virðan* (pp.) *mik létu* with which they caused me to be honoured 67/6.

virðandi *m.* valuer, one who sets a (high) price on s-thing, in kenning for ruler (Skúli): *v. liðar skerja* 46/1.

virðar *m. pl.* men 37/1 (dat. with *banna*), 46/1 (dat. with *gefr*), 64/1 (dat. with *mœtti*).

vísa *f.* strophe, stanza (= **eyrindi, ørindi**) 1/20, 27, 2/9, 6/19, 7/9, 8/12, 16, 28, 31, 15/9, 11, 16/9, 15, 17/26, 51/10, 53/14, 58/15, 81/9.

vísi *m.* leader, ruler; = Hákon 5/7, 16/1, 20/7 (obj. of *lýsa* or gen. with *iðjur*); = Skúli 38/3 (gen. sg.), 52/5, 62/5 (subj. of *heldr*—or **skiptir*, see note), 72/7, 79/1, 93/3.

vísuhelmingr *m.* half-stanza, quatrain (= **helmingr**) 8/25, 27, 33, 15/9, 11, 39/10, 48/9. Cf. *FoGT* 136, 140, 145.

vísulengð *f.* length of a verse; *of alla v.* throughout the verse 6/14, *v. saman* throughout the verse 43/9.

vísuorð *n.* line (of a verse) 0/20, 22, 1/11 &c.; *í fyrsta (fyrra) vísuorði* in the first line of (each) couplet 1/13, 28; *í ǫðru vísuorði* in alternate lines, in the second of each pair of lines 1/12, 17/35; *í vísuorði* in a line 43/11; *eptir v.* at the end of a line 59/10. Cf. *TGT* 63–5, 83, 92, *FoGT* 137–8.

vita (veit, vissa) *pret.-pres. vb.* know 62/2; with *at*-clause 90/7; *veit ek* as parenthesis or with *hvar*-clause 42/3; with acc. and inf. 74/5 (subject is *buðlungr*), 86/1; with acc. and a., know s-one is s-thing 69/3, with acc. (*unga jǫfra*) and comp. a. 98/1; subjunc. with acc. (*mærð*) and pp. (*orta*) 95/5; with suffixed pron. (*vissak*), acc. and sup. adv. ('whom I knew [to be] quite the most outstanding') 67/2.

viti *m.* beacon, fire in kenning for gold (as ornament for arms): *v. valstaða* 86/7 (subj. of *vaða*).

vitni *n.* witness, testimony 18/10. Cf. **vitnir**.

vitnir *m.* 'watcher', wolf 8/36 (i.e. Fenrir, in kenning for Óðinn), 9/4, 11/5 (dat.), 18/2 (obj. of *fremr*), 18/15, 56/6 (dat. with *gefr*).

vitr *a.* wise 16/1.

vægðarlauss *a.* merciless 65/2 (with *hríð*).

væni *n.* expectation; *er e-m v. til e-s* s-one can expect s-thing 58/2.

vǫllr *m.* field, battle-field; dat. sg. *velli* 10/5; pl. *vellir* 9/1.

vǫndr *m.* wand in kennings for sword 6/15, *vígfoldar v.* (in kenning for warrior) 30/6.

vǫr *f.* landing-place 20/8 (also by word-play 3rd pers. sg. of *vara* wv. 'warn').

vǫrðr *m.* defender, guardian; *grundar v.* = ruler 90/8 (subject of *gata*: 'I know that no ruler has ever been gifted with . . .').

yggr *a.* frightening 7/5 textual note (cf. *LP*).

ýgr *a.* terrible, fierce 7/5.

ylgr *f.* she-wolf 51/7, 56/2 (obj. of *ali*), 66/4, 96/3 (gen. with *granar*).

ýmiss (yms-) *a.* various 58/15.

yngvi *m.* ruler, = Skúli 62/3, 93/7. Cf. *Skáldskaparmál*, ch. 64.

yppa (t) *wv.* with dat. raise; make known 80/7.

ýr *m.* (yew-)bow 16/3.

yrkja (orta, pp. **ortr)** *wv.* compose (poetry) 0/2, 8/28, 18/13, 35/12, 41/10, 43/9, 44/10, 51/10, 53/12; with acc., compose in (verse-forms) 100/6 (pp.);

with dat., using s-thing 17/35, 19/10; *y. at e-u* compose with, using s-thing 1/53; *y. eptir* compose in, in accordance with (a metre or verse-form) 44/10, 67/13, 14, compose in imitation of 58/15; *y. með* compose using (verse-forms) 1/42 (pp.), 53/14, 67/1 (p. with suffixed pron.), 83/9, *hvar viti maðr mærð orta með* where would one know of praise composed in 95/5; *ort er* it (*lof*) has been composed 96/1.

ýskelfir *m.* (yew-)bow-shaker, warrior, = Hákon 11/1.

ýta (tt) *wv.* push; pres. p. *ýtandi auðs* distributor of gold, generous ruler, = Skúli 46/5.

ýtar *m. pl.* men 41/1 (dat. of advantage with *velr*), 42/7 (dat. of respect with *arm*, which is collective), 46/5 (acc.), 63/6 (gen. with *ferðar*), 97/3.

þá (1) *adv.* then 72/7; *þá er* conj. when 36/1, 39/5, 55/7, 75/3, 88/1, 95/2, 101/4, 5.

þá (2) *p.* of **þiggja** 27/5.

þá (3) *acc. pl.* of **sá (2)**; *þá er* (rel.) 67/2 (with *siklinga*), 98/2 (with *jǫfra*).

þakka (að) *wv.* thank (*e-m e-t* s-one for s-thing) 46/8; with suffixed pron. *-k* 28/2.

þar *adv.* there 16/2; belonging in following *at*-clause 58/3 (cf. **svá**); *þar er* conj., where 45/7, 46/7.

þat *conj.* = *at* ?92/4 (with *þess*), see *NN* 3263.

þáttr *m.* strand; *við þátt* in or with strand(s) 89/8 (with *slungit*).

þegja (þagða) *wv.* be silent 85/6.

þegn *m.* subject 42/6 (dat. of respect or advantage with *fit*).

þekkr *a.* pleasing, welcome 87/4 (with *bekkr*).

þelli *n.* firs (collective); by synecdoche for ship(s) 20/2.

þengill *m.* king, ruler; = Hákon 4/6, 5/4, 21/1 (*lýða þ.*); = Skúli 31/4, 37/7 (*þ. skatna*), 69/3, 89/3, 92/3, 93/1. Cf. *Skáldskaparmál*, ch. 64.

þiggja (þá) *sv.* receive 27/5, 49/7; *þ. at e-m* receive from s-one 91/1; get, gain 59/7.

þilja *f.* plank, planking 22/2 (obj. of *strýkr*); in kenning for shield: *þ. Hrungnis ilja* (part of kenning for man) 30/4.

þing *n.* assembly = battle 33/3, 36/8 n. (subject of *stóð*; cf. **egg**); in kenning for war-leader (Skúli): *þings þrøngvir* 41/3.

þinga (að) *wv.* adjudge, settle, determine; *þ. brott e-u* give s-thing away 47/4.

þjóð *f.* people 21/5, 66/8 (dat. with *refsa*), 69/2 (dat. with *kunn*), 88/7, 90/1; company, group (with *hǫlða áttar*) 37/8 (indirect obj. of *bjóða*; cf. n.).

þjóðá *f.* great river; *þjóðár hræs*, i.e. rivers of blood 7/4.

þjóðkonungr *m.* king of a nation, great king, = Hákon 12/4, 11, 97/2.

þjóðsterkr *a.* mighty strong 33/4 (with *stǫng*).

þjóta (þaut) *sv.* emit a noise, resound 34/7.
þollr *m.* fir; in kennings for warrior(s): *þ. skjalda* 30/1 (= Snorri), *þ. jǫru* 53/8. Cf. **fleinþollr**.
þora (ð) *wv.* aux. with inf. dare 8/7, 23.
þorna (að) *wv.* become dry 24/2.
þrábarn *n.* beloved (longed for) child (offspring); demanding child? 32/8 (dat. of respect with *fet*).
þrár *a.* obstinate; n. as adv., irresistibly 33/3, powerfully or repeatedly 37/5, constantly 89/3.
þrekr *m.* endurance, fortitude 15/6.
þrennr *a.* triple, having three parts or divisions 0/5, etc.; pl. in sets of three 36/9; in pl. can mean simply 'three' 69/4 (with *kvæði*: the poem is in three parts or there are three poems, i.e. three series of stanzas; the reference seems to be to the second section of *Háttatal* and two other poems, so that the third section will make a fourth poem about Skúli, cf. 69/5 and 95/3, and see Finnur Jónsson 1920–4, ii. 78; *Sturl.* i. 278).
þrífask (þreifsk) *sv. md.* thrive, go forward, rage 32/1.
þríhendr *a.* triple-rhymed; n. as subst. 35/14, 36/1 textual n.
þrima *f.* noise, uproar, battle 9/2.
þrjóta (þraut) *sv.* impers. with acc., one stops; with *at* and inf. 31/6; *e-n þrýtr e-t* one comes to lack s-thing; pp. *þrotinn*: *spyr ek gotna þrotna elli* I hear men [are] deprived of (have lost) old age, i.e. they die young in battle 61/8 (alternatively *þrotna* may be inf. here).
þróask (að) *wv. md.* grow, thrive, increase 4/1, 10, 11.
þroski *m.* development, advancement 30/3.
þrotna (að) *wv.* come to an end, cease to be; *spyr ek gotna elli þrotna* I hear that men's old age is being prevented ?61/8, cf. **þrjóta**.
þruma *m.* noise; in kenning for battle, *þ. randa* 8/8 textual n.
þryngva (þrǫng) *sv.* press, proceed energetically (*at* to) 16/7.
þrǫmr *m.* edge, side (of shield) 8/8 n., 54/4, 61/4 (obj. of *nemr*); side of ship, gunwale 34/7, 38/6 (dat. obj. of *svipa*, or adv., 'with its fine sides', if *svipa* is intrans.); pl., coast 16/8 (obj. of *stikar*).
þrǫngr *a.* narrow 21/6; n. as adv., crowded, in large numbers 88/7; (filled) tight 89/2.
þrøngvir *m.* presser. compeller; in kenning for war-leader: *þings þ.* (= Skúli) 41/3.
þungfarmr *m.* heavy burden; in kenning for gold: *þ. Grana* 41/4 (see *Skáldskaparmál*, ch. 40).
þunghúfaðr *a.* (*pp.*) heavily planked; sup. 34/8.
þungr *a.* heavy 21/5, 46/4 (with *vellum*), 74/8 (with *víði*); harsh 64/6.
þunnr *a.* thin, slender 7/3, 19/7.
þverra (ð) *wv.* (cause to) decrease; *þ. auð* be generous 40/4.
þykkja (þótta) *wv.* be thought 6/13, 16, 8/27, 44/9, 58/16, 94/7.

þyrja (þurða) *wv.* rush along 6/7 (with acc. of route); rage, whistle past 20/7, 59/6.

þyrna (d) *wv.* spike, hedge, surround, or enclose with s-thing sharp (with instrumental dat.) 79/5.

þǫgn *f.* silence 25/3 (obj. of *fellir*; i.e. makes men loquacious).

æ *adv.* for ever 96/5.

ægir *m.* sea, ocean 3/1 (in kenning for gold), 102/6.

ætt *f.* family line; people 26/4; direction, region; *or ættum* out of proportion, beyond bounds, off course, excessive 8/30. Cf. **átt**.

ættstuðill *m.* pillar of family, outstanding member of dynasty; *æ. skylja* pillar of the line of kings 2/4.

œðri *a. comp.* higher, better, greater, nobler 90/7 (with *gjǫflund*), 91/7 (with *mann*), 95/6 (with *hætti*).

œgir *m.* terrifier; in kenning for ruler (= Skúli): *œ. jǫfra* 55/1 (dat. of comparison with *betra*).

ǫðlingr *m.* nobleman, prince (= Hákon) 25/1.

ǫflugr *a.* mighty 65/7.

ǫl *n.* ale 23/5, 25/2.

ǫld *f.* mankind, men 7/6 (obj. of *lætr*), 13/2 (with *ǫll*, subject of *lýtr*), 14/7, 50/2 (dat. of respect with *tjǫld*), 61/4 (gen. with *skjaldar*), 67/1 (dat. of advantage with *minnum*), 82/3, 96/6; pl. *aldir* 16/4 (subject of *hníga*), 34/2, 43/2 (obj. of *fremr*); in kennings for ruler (Skúli): *aldar hǫfundr* 37/2, *aldar vinr* 54/7.

ǫlsaðr *a.* ale-sated 86/4 ('so as to make them filled with ale').

ǫndurðr *a.* beginning of, early part of 57/10.

ǫr (pl. **-var**) *f.* arrow 9/8.

ǫrbrjótr *m.* eager breaker; in kenning for warrior (Snorri): *odds bláferla ǫ.* eager breaker of shields 31/6.

ǫrn *m.* eagle 32/8, 51/3.

ǫrr *a.* liberal (*e-s* with s-thing); giving rise to a great deal of s-thing 80/6 (with *fǫrum*).

ǫxl *f.* shoulder 2/2.

ørindi = **eyrindi** 8/48.

ørverðr *a.* unworthy (*e-s* of s-thing) 100/4.

Index of Names

General Index

9 780198 112389

Printed by Libri Plureos GmbH in Hamburg,
Germany